手把手教你做一款商用智能电话定位手表

MTK 智能穿戴开发实战

何润平　刘　燃　编著

西安电子科技大学出版社

内 容 简 介

MTK 是当下最流行的智能穿戴设备首选研发平台。本书以"MTK6260 儿童定位智能电话手表"和"MTK2502 GPS 防丢追踪器"为例，按照产品开发流程，由浅入深地带领读者快速掌握 MTK 智能穿戴开发的所有技能。本书内容包含了 MTK 智能穿戴开发所需的各方面技术知识，从获取开发工具、搭建开发环境、MMI 编程、网络与定位到真机下载与调试，都有详细讲解。

对于想要从事 MTK 智能穿戴产品研发工作的在校学生、程序开发爱好者或转行从业者，本书是一本很好的入门教材；而对于已经入行，正在从事 MTK 智能穿戴产品软件开发的程序员来说，本书也能给予一定的参考和指导。本书语言通俗易懂，即使是从没接触过 MTK 开发的读者也能顺利上手，并能根据书中的实例展开实践。

图书在版编目(CIP)数据

MTK 智能穿戴开发实践 / 何润平，刘燃编著. —西安：西安电子科技大学出版社，2018.1
ISBN 978-7-5606-4824-8

Ⅰ. ① M… Ⅱ. ① 何… ② 刘… Ⅲ. ① 移动电话机—操作系统 Ⅳ. ① TP316

中国版本图书馆 CIP 数据核字(2018)第 002461 号

策　　划	高　樱
责任编辑	黄　菡　阎　彬
出版发行	西安电子科技大学出版社(西安市太白南路 2 号)
电　　话	(029)88242885　88201467　　　邮　编　710071
网　　址	www.xduph.com　　　　电子邮箱　xdupfxb001@163.com
经　　销	新华书店
印刷单位	陕西华沐印刷科技有限责任公司
版　　次	2018 年 1 月第 1 版　　2018 年 1 月第 1 次印刷
开　　本	787 毫米×1092 毫米　　1/16　　印 张　16
字　　数	380 千字
印　　数	1～3000 册
定　　价	59.00 元

ISBN 978-7-5606-4824-8 / TP

XDUP 5126001-1

如有印装问题可调换

前言

MTK(Media Tek Inc)是台湾联发科技股份有限公司的简称，它是全球著名的 IC 芯片设计公司，发布了许多低成本、低功耗、小体积的芯片整合方案，这些方案被广泛用于无线通信及数字多媒体领域的电子产品开发中，比如手机、路由器等。随着物联网的快速发展，MTK 也适应时代发展潮流，推出了越来越多的衍生产品，比如共享单车的智能锁、智能电话手表、POS 机、点读机、机器人、防丢防盗设备等都选用了 MTK 平台方案。MTK 的产品以开发周期短、成本低、功耗低、可扩展性强等优点被越来越多的公司企业所认可。

MTK6260、MTK2502、MTK2503 等平台是当下最流行的智能穿戴设备首选研发平台，比如华为和 360 的儿童手表、步步高的小天才和糖猫儿童电话手表等，都是基于 MTK 功能机平台研发的。虽然 MTK 平台应用如此广泛，但是对于想要从事 MTK 开发的人员来说，却连入门都有点困难。因为 MTK 是一个半封闭的系统，市场上关于它的书籍、教程等资料非常少，联发科的官方资料也需要付费才能获取，而且这个费用对于个人开发者是难以承受的，这就导致想要从事 MTK 平台研发的人员不得其门而入，而且随着近几年 Android、iOS 等移动开发的流行，导致了解 MTK 平台、从事 MTK 平台研发的人员越来越少。正是基于此种现状，作者决定撰写本书，希望根据自己多年的 MTK 研发经验，并以"MTK6260 儿童定位智能电话手表"和"MTK2502 GPS 防丢追踪器"为例，讲解实际产品开发流程，总结实际项目开发中的常见问题及常用知识点，帮助读者快速入门并学会 MTK 智能穿戴开发技能。

本书的内容几乎涵盖了 MTK 智能穿戴设备软件开发的所有知识点，有些知识点讲得并不是很深入，但作者给出了如何获取相关资料的途径。书中的章节内容都是根据实际项目开发步骤，按照从易到难的顺序排列的，建议读者按顺序学习。前面两个章节是 MTK 平台相关的基础知识，读者首先需掌握开发环境的配置，然后掌握系统的编译方法。只有配置好了开发环境并能使用 make new 指令编译代码通过，才能进行后面章节的学习。在学习完所有的知识点后，作者以一个游戏实例来提高读者的学习兴趣，让读者学会如何运用前面所学的知识点。最后本书还介绍了两个商用的智能穿戴产品作为读者实战开发的调试设备。

本书的特点如下：

(1) 实用性强。以真实的商用产品"儿童定位智能电话手表"和"GPS 防丢追踪器"为例，全面讲解 MTK 智能穿戴产品的开发流程和技能。

(2) 专业权威。作者是 MTK 智能穿戴领域的一线开发者，拥有多年 MTK 项目开发经验，负责多款智能穿戴产品的开发及量产维护工作，书中内容全部来自真实项目的开发总结。

(3) 内容全面。本书内容基本涵盖了 MTK 智能穿戴设备软件开发的所有知识点。

(4) 实验可靠。书中所有源码都经过真实环境验证，有极高的含金量。

(5) 售后答疑。读者可在 https://www.fengke.club/GeekMart/su_f7iGGsI44.jsp 官网社区提问，笔者会不定期答疑。

本书的适用范围如下：

(1) 希望从事 MTK 研发工作的在校学生、程序开发爱好者或转行从业者。

(2) 已经入行，正在从事 MTK 智能穿戴产品开发的工程师。

(3) MTK 智能穿戴技术的培训机构和单位。

(4) 高校教师或学生，本书可作为高校实验课程教材。

本书第一章和第七章由刘燃、何润平共同编写，其他章节由何润平编写。特别感谢深圳疯壳的各位朋友，他们对本书的编写提供了可靠的技术支撑与精神鼓励。此外，还要感谢西安电子科技大学出版社的相关工作人员，正是他们的辛勤劳动才使本书得以顺利出版。

关于本书的源码、视频套件等，读者可以通过 https://www.fengke.club/GeekMart/su_f7iGGsI44.jsp 社区论坛免费下载。由于时间仓促及作者水平有限，书中难免存在不足之处和纰漏，恳请读者批评指正，可通过社区论坛与作者互动。

作　者

2017 年 11 月

目 录

第一章 开发准备 1
1.1 平台简介 1
1.2 开发套件 2
1.3 开发环境搭建 3
1.3.1 安装 RVCT 4
1.3.2 安装 ActivePerl 16
1.3.3 安装 Office 17
1.3.4 安装 USB 驱动 17
1.3.5 环境检测 18
1.3.6 编译环境错误分析 18
1.3.7 其他工具软件 20

第二章 开发基础 21
2.1 编译指令 21
2.2 系统框架 26
2.3 新增功能模块 29
2.4 第一个程序 44
2.5 下载和调试 53
2.5.1 FlashTool 下载 53
2.5.2 Catcher 调试 60

第三章 MMI 基础编程 72
3.1 资源 72
3.1.1 新增".res"资源文件 72
3.1.2 字符串资源 75
3.1.3 屏幕资源 80
3.1.4 图片资源 83
3.1.5 菜单资源 86
3.1.6 铃声资源 93
3.1.7 NVRAM 资源 94
3.1.8 定时器资源 95
3.1.9 消息资源 96
3.2 绘制界面 99
3.2.1 清屏 100
3.2.2 显示图片 101

	3.2.3 界面排版	104
	3.2.4 绘制几何图形	105
	3.2.5 给界面添加背景音乐	107
	3.2.6 定制屏幕尺寸	109
3.3	按键	110

第四章 MMI 高级编程 ... 112

4.1	定时器	112
	4.1.1 普通定时器	112
	4.1.2 Reminder 定时器	116
4.2	层	121
4.3	文件管理	125
	4.3.1 目录管理	125
	4.3.2 文件操作	127
4.4	NVRAM	130
	4.4.1 存储简单数据的 NVRAM	131
	4.4.2 存储复合数据的 NVRAM	133

第五章 网络与定位 ... 139

5.1	SIM 卡通信	139
	5.1.1 短信	139
	5.1.2 通话	141
5.2	Socket 网络编程	143
5.3	网络通信协议	174
	5.3.1 网络连接协议	174
	5.3.2 数据传输协议	176
5.4	定位	182
	5.4.1 GPS 定位	182
	5.4.2 LBS 定位	197

第六章 游戏开发 ... 205

6.1	触屏	205
6.2	铃声播放	207
6.3	游戏说明	208

第七章 项目实战 ... 212

7.1	儿童定位智能电话手表	212
7.2	GPS 防丢追踪器	226

附录 A	Source Insight 工具介绍	237
附录 B	Beyond Compare 工具介绍	245
参考文献		250

第一章 开 发 准 备

1.1 平 台 简 介

MTK 是台湾联发科技股份有限公司(MediaTek Inc)的简称，它是全球著名的 IC 设计厂商，专注于无线通信及数字多媒体等技术领域，其提供的芯片整合系统解决方案包含无线通信、高清数字电视等相关产品。

在物联网高速发展的今天，采用 MTK 方案定制的产品已经无处不在，它们在给我们提供娱乐的同时，也在慢慢地改变我们的生活，其中大家最熟悉的就是手机，许多知名手机开发商都选用过 MTK 平台方案，包括华为、小米、酷派、vivo、OPPO、三星等。

智能手机在 2010 年开始风靡市场，但在其出现之前，市场上 70%的手机都采用 MTK 平台方案，它有个不好听的名字叫"山寨机"，业内人士则称之为"功能机(feature phone)"。虽然现在"功能机"已经慢慢地淡出市场，但"功能机"曾经的辉煌绝对远超现在的智能机，而且"功能机"的系统并没有被淘汰，只是换了一种产品形态出现，比如智能电话手表、智能手环、宠物跟踪器、机器人、无人机以及共享单车和共享雨伞上的智能锁等。这些产品都是使用 MTK 功能机系统研发的，并以低功耗、低成本、较高的运行效率等优势，依然在行业内具有不可替代性，几乎每隔一段时间都会有一个新的 MTK 产品问世。而且 MTK 功能机平台也在不断地发展，它的未来只会更加辉煌。

虽然 MTK 的产品非常普遍，但对于想学习 MTK 产品研发的人，入门却有点难度。因为 MTK 的系统是半封闭的，所有的官方资料都是要付费获取的，这对于初学者个人而言显然门槛有点高，尽管网上可以找到一些资料，但全都是零零散散的只言片语，没有全面系统的内容介绍。鉴于此种情况，为了帮助更多的人学习并掌握 MTK 软件开发，笔者把自己多年的项目开发实践经验总结出来，整理成本书。本书为大家讲解从事 MTK 功能机软件研发岗位所需的知识及技巧，并在最后的章节中，运用前面所学的知识开发一款商业化的儿童定位智能手表。只要充分掌握本书的知识，不管是从事功能手机的研发工作，还是智能穿戴设备的研发工作，或者是物联网产品模块的研发工作，都能够胜任。有了扎实的 C 语言基础和 MTK 的软件开发经验，也能够从事 MTK 同类平台(比如展讯)的软件研发工作。

MTK 平台的代码是半开源的，我们可以看到大部分的源码，这些源码几乎全部都是采用纯 C 语言开发的，所以在学习本书的时候，C 语言是读者必须掌握的程序设计语言。只要具有扎实的 C 语言基础知识，很多知识点都可以通过阅读源码来学习。所以本书不仅向读者阐述相关知识点，也会讲解如何看 MTK 平台上的原生代码，以培养读者的自学能力。至于代码编写风格，大家都知道，C 语言是面向过程的程序设计语言，其程序的组成单位是函数。在写 C 语言程序时，其代码的管理是非常重要的。一个好的 C 语言程序，不仅要实现相应的功能，而且要注重代码的可维护性和可读性，做到"高内聚，低耦合"的软件设计标准。所以在进入实际编程之前，应先学会如何创建一个自己的功能模块，通过创建自己的功能模块来管理自己编写的代码。功能模块里面包含了很多子功能，比如 Socket 联网、定位等，这些子功能都是经过封装的，只提供一些 API 供其他功能或其他软件开发人员调用，其代码可通过宏来包含并定义对应宏开关。在 MTK 平台上进行软件开发，读者应适应并学会这种代码的管理方式，而且尽量把一些功能模块化，这样可以实现代码的复用，提高开发效率，并形成自己的代码库。

1.2 开发套件

本书中提到的所有工具软件以及源代码都可以在 https://www.fengke.club/GeekMart/su_f7iGGsI44.jsp 官网社区下载，本书配套的 MTK 硬件开发套件、配套视频教学课程也可通过该网站获取。在学习的过程中如果碰到问题，可直接在官网社区提问，笔者会在里面为各位读者答疑。

本书涉及三套实验源码，分别是 MTK6260A、MTK6260M、MTK2502A。在项目实战篇之前，所有章节实验都是基于 MTK6260A 源码修改讲解的，这部分实验可以直接在模拟器上运行，但是也有一些功能现象必须在硬件平台上才能更好地体现，比如定位。此外，有些实验在模拟器上执行与真机上执行会稍有差异，比如网络编程部分，模拟器上网是基于 PC 网卡，而真机上网是基于 SIM 卡，所以学习本书的时候，建议读者自备一套 MTK 硬件开发套件，同时准备一张支持 2G 网络的 SIM 卡(有的 4G 卡不兼容 2G，需向运营商确认)。在项目实战篇，我们以"儿童定位智能电话手表"和"GPS 防丢追踪器"这两个商用项目为例，讲解在真机环境下 MTK 编程开发方法。根据项目不同，会分别使用到 MTK6260M 实验源码和 MTK2502A 实验源码。其中："儿童定位智能电话手表"外观如图 1.2-1 所示，选用的是 MTK6260M 平台；"GPS 防丢追踪器"外观如图 1.2-2 所示，选用的是 MTK2502A 平台。项目实战篇所有的实验现象，都是下载到真机(也就是文中所说的 MTK 硬件开发套件)上直接运行的。

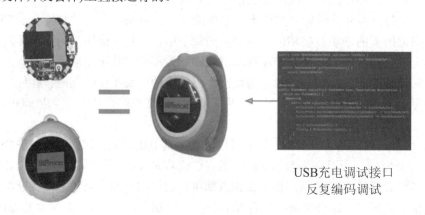

图 1.2-1

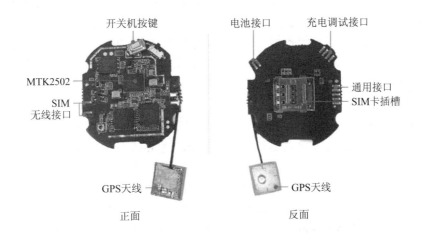

图 1.2-2

在使用本书配套实验源码的时候，需要特别注意项目名称的差异。在实战篇之前，所有章节的实验都是基于 MTK6260A 源码修改的，其项目名称为"HEXING60A_11B"，"儿童定位智能电话手表"配套源码的项目名称为"UMEOX60M_11B"，"GPS 防丢追踪器"配套源码的项目名称为"HEXING02A_WT_11C"。在使用 make 指令编译的时候需要注意项目名称的切换，在后面讲解"编译指令"的章节中，会告诉大家如何查找任意一份源码的项目名称。后续也许会给读者提供更多的开发套件，每个开发套件配套的源码也可能是不同的，读者可在官网社区上获取相关信息。

在使用本书配套开发套件的时候，需要注意，我们提供给读者的 MTK 硬件开发套件都是商用产品的雏形，为了避免一些商业纠纷，也许部分型号的设备不会以最终的产品形态提供，但对于学习开发完全够用。另外，每个产品的需求定义都不同，有的产品不带屏幕或者屏幕很小，又或者屏幕上显示内容的方式与 MTK 的函数接口不兼容，在这些产品上都无法演示书中的屏幕显示部分代码。笔者的建议是，能够在模拟器上完成的工作，尽量在模拟器上完成，尤其是关于界面显示部分的内容，使用模拟器可以大大提高我们的开发效率。模拟器配合 Catcher 工具可以模拟大部分的真机环境，当模拟成功了，我们再把代码烧录到硬件设备中。如果无法通过屏幕看到现象，可以使用 Catcher 工具通过打 trace(打 log)的方式查看代码运行的情况。这些内容都会在书中一一介绍。还有一点需要特别注意，每个设备都有一个 IMEI 号码，一般会印在包装盒上，在后面做 Socket 网络测试连接服务器的时候，这个 IMEI 号码是服务器识别终端的唯一标志，所以千万不能把这个号码弄丢了。

最后，大家需要明确的是：MTK 的软件都是相通的，不同平台之间的代码大同小异，平台的差异主要是针对硬件配置。本书的实例代码，在 MTK 的任意平台上都可以运行，并且效果不会有很大差别。

1.3 开发环境搭建

MTK 在 10A(包含)以后的软件版本都使用 RVCT 编译工具。RVCT 是 RVDS 的一个组件(编译工具链)，系统中可以单独安装 RVCT，也可以和 10A 以前的版本的编译工具 ADS 共存，但 ADS 对我们已经没用了。

现在市场上主流的 MTK 版本都是比 10A 更新的版本，比如 MT6260、MT6261、MT2502、MT2503 等，这些版本的编译环境都是 RVCT，目前比 10A 老的版本，比如 MT6225、MT6235 等基本上已经淘汰了，所以本书的编译环境只介绍 RVCT 的安装方法。

安装 RVCT 的电脑基本配置如下：

(1) CPU：目前市场上主流的 CPU 都可以，但推荐使用 Intel 酷睿系列 CPU，不建议使用 AMD 速龙系列的 CPU，笔者尝试过在多台 AMD 速龙 CPU 的电脑上安装 RCVT 都无法成功。

(2) 内存：至少 2G(内存越大，编译速度越快)。

(3) 操作系统：RVCT 完美支持 Windows XP 系统，但目前 Windows XP 系统用户较少，本书介绍的安装环境为 Windows 7 64 位旗舰版操作系统。至于比 Windows 7 更高版本的系统 Windows 8、Windows 10，笔者也尝试过，但无法安装成功。有兴趣的读者可以自己尝试研究，但在学习本书内容时，建议读者使用与本书匹配的操作系统——Windows 7 64 位旗舰版。

1.3.1 安装 RVCT

为了防止杀毒软件把环境安装包里面的某些文件当做病毒处理，我们首先关闭所有杀毒软件以及 360、腾讯等电脑管家类的软件，然后安装 RVCT，具体步骤如下：

(1) 鼠标右键单击电脑桌面空白处，选择"控制面板"→"外观和个性化"→"个性化"，把"基本和高对比度主题"改为"Windows 经典"，如图 1.3-1 所示。

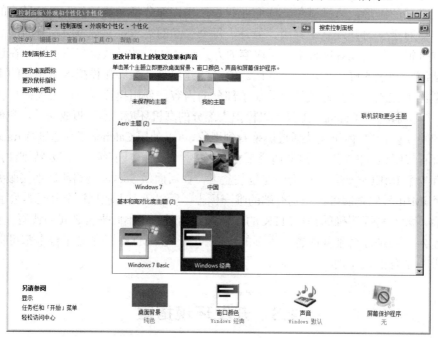

图 1.3-1

(2) 鼠标右键单击桌面图标"计算机"，选择"管理"，依次单击"服务和应用程序"→"服务"，在中间窗口中找到名称为"Themes"的服务，单击"停止此服务"，将该服务停止，如图 1.3-2 所示。

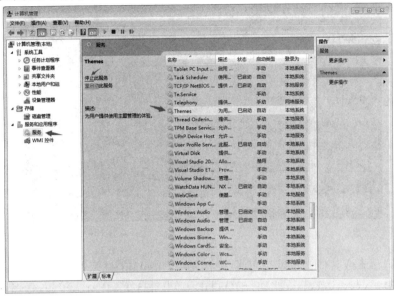

图 1.3-2

(3) 进入 C 盘(系统盘)，在根目录下用鼠标右键单击 Program Files 文件夹，选择"属性"，在弹出的属性窗口中去掉"只读"属性，然后连续两次单击"确定"按钮。大多数电脑的 Windows 7 系统都可以不执行这一步，但为了在后面安装 RVCT 的时候防止 License 的写权限被限制，我们还是把它加上，如图 1.3-3 所示。

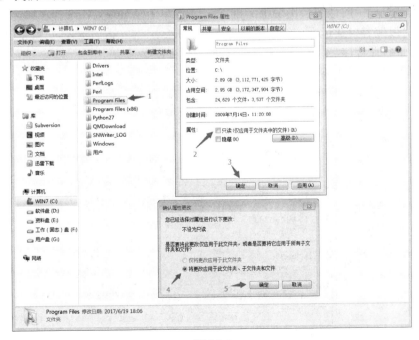

图 1.3-3

(4) 单击"开始"(电脑桌面右下角按钮)，输入"cmd"，按回车键，进入DOS界面，输入"ipconfig -all"，找到本机网卡的物理地址，如图1.3-4所示。

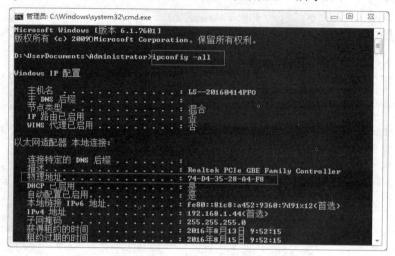

图1.3-4

然后用记事本或其他文本编辑工具(推荐使用UltraEdit)打开RVCT 3.1\rvds4cr\rvds.dat，找到HOSTID = xxxxxxxxxxxx(比如：HOSTID = 74D43528A4F8)，将xxxxxxxxxxxx替换为本机网卡物理地址(比如：图1.3-4所示网卡物理地址为74-D4-35-28-A4-F8)，总共需要替换19处，然后保存、关闭。

(5) 进入RVCT 3.1\RVCT31build569\RVDS_3_1目录(如果RVCT的完整目录包含中文，则在后面安装License的时候，有的操作系统会报错，但有的系统不会报错。如果想要避免这个问题，可以将RVCT不要放在中文目录下，笔者的系统在中文目录下安装不会有问题，所以下面安装过程中的截图会出现中文目录)，右键单击setup.exe，选择"属性"→"兼容性"，勾选"以兼容模式运行这个程序"兼容"Windows XP(Service pack 3)"，并勾选"以管理员身份运行此程序"，单击"确定"按钮，如图1.3-5所示。

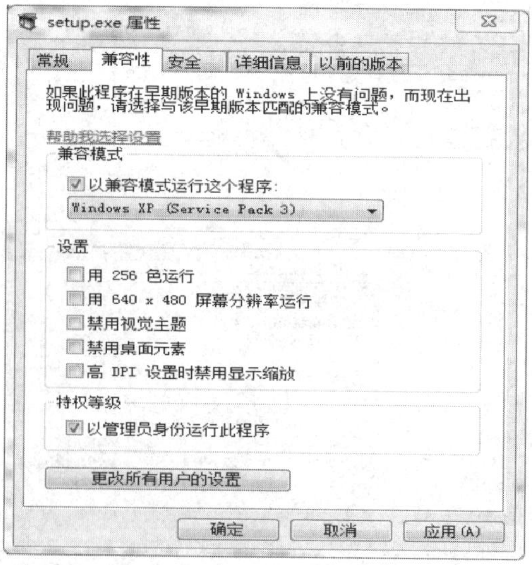

图1.3-5

(6) 双击打开 setup.exe 文件,开始安装,此时首先会显示一个 DOS 窗口,如图 1.3-6 所示。

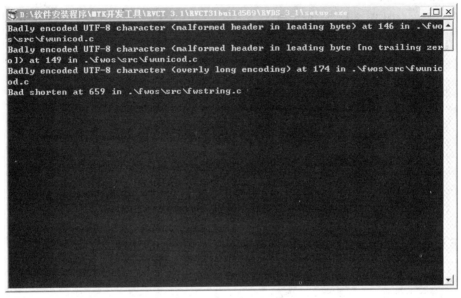

图 1.3-6

过几秒后,当出现图 1.3-7 所示界面时,则说明成功启动安装程序;如果没有出现该界面,则查看下面的"编译环境错误分析",依次单击"下一个"按钮,如图 1.3-8、图 1.3-9 所示。

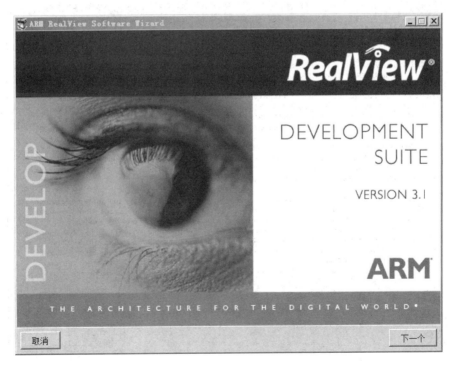

图 1.3-7

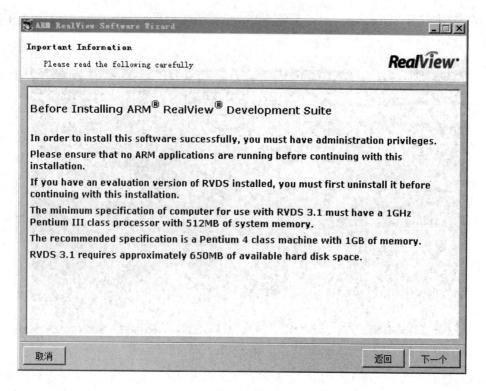

图 1.3-8

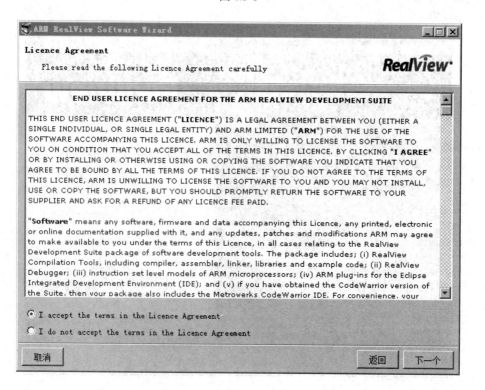

图 1.3-9

(7) 当出现安装路径选择时,务必选择安装在默认目录 C:\Program Files 下(如果没有安装在默认目录,则编译时需要在 MTK 系统文件(option.mak)中修改代码配置,这样比较麻烦,不建议这样做),然后单击"下一个"按钮,如图 1.3-10 所示。

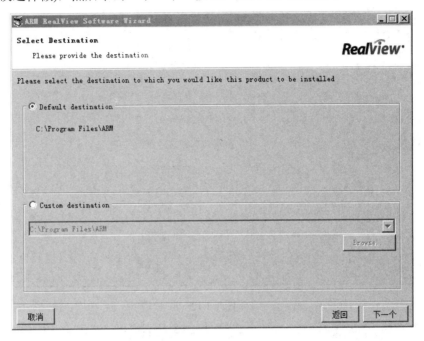

图 1.3-10

(8) 当出现图 1.3-11 所示界面时,选择"RVCT Only",然后依次单击"下一个"按钮,如图 1.3-12、图 1.3-13 所示。

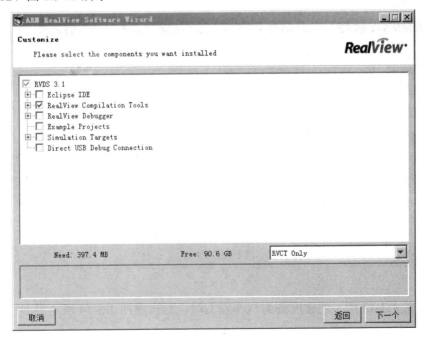

图 1.3-11

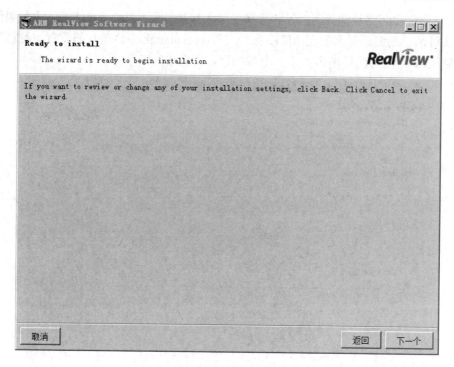

图 1.3-12

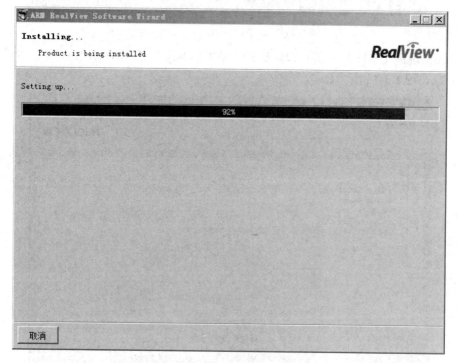

图 1.3-13

(9) 当图 1.3-13 所示的进度条到 90% 左右时，会出现图 1.3-14 所示界面，单击"下一步"按钮。

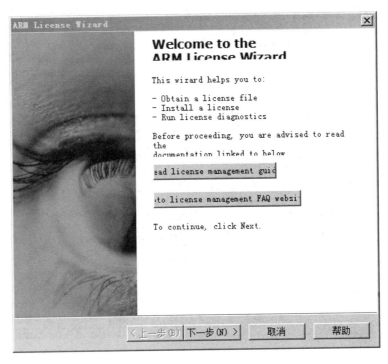

图 1.3-14

(10) 如图 1.3-15 所示,选择中间的"Install license",然后依次单击"下一步"按钮,如图 1.3-16 所示。

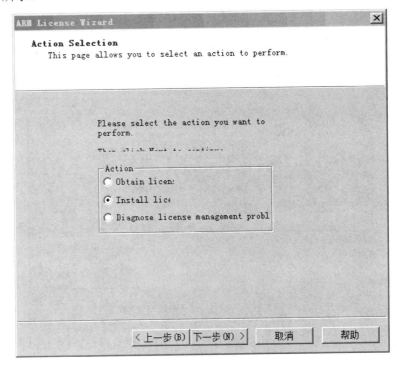

图 1.3-15

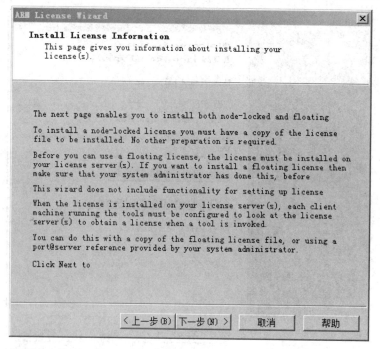

图 1.3-16

(11) 在如图 1.3-17 所示界面中,单击"..."按钮,选择第(4)步中修改的 RVCT 3.1\rvds4cr\rvds.dat 文件(笔者把 rvds4cr 文件夹拷贝到了 D 盘根目录下),然后单击"Add"按钮,如图 1.3-18 所示,在弹出的提示框中都选择"是",如图 1.3-19 所示。在点击"Add"之前,要确保 rvds.dat 文件没有被其他进程占用,并且最好不要包含在中文目录中。

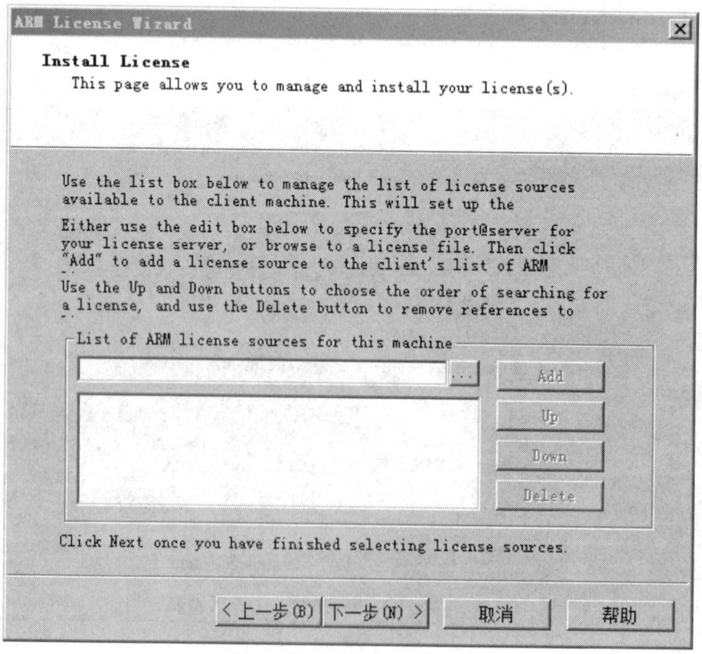

图 1.3-17

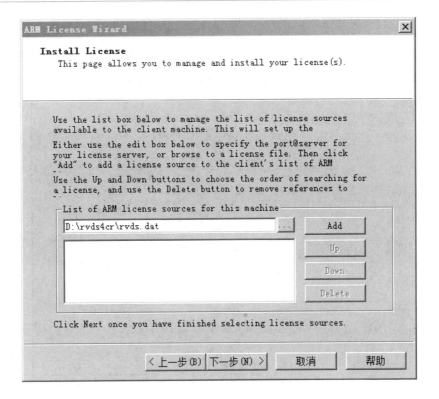

图 1.3-18

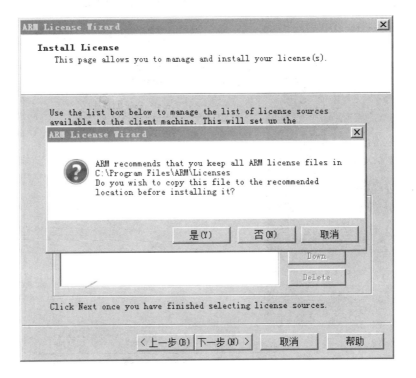

图 1.3-19

(12) 最终执行结果如图 1.3-20 所示，即 License 已添加成功，然后单击"下一步"按钮。

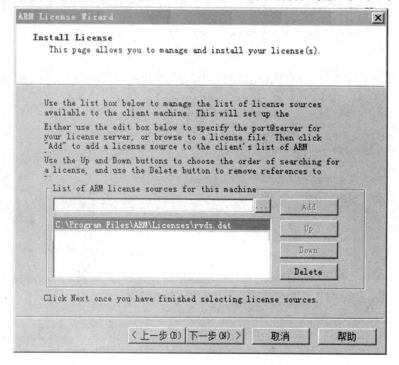

图 1.3-20

(13) 如图 1.3-21 所示，License 安装完成，单击"完成"按钮。

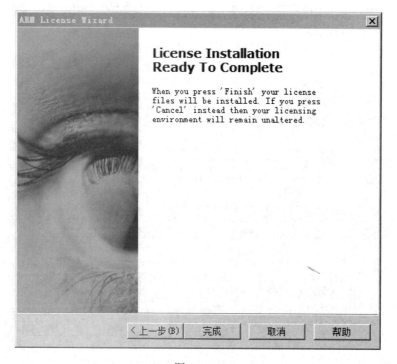

图 1.3-21

(14) License 安装完成之后，进度条走到 100%，单击"下一个"按钮，如图 1.3-22 所示。

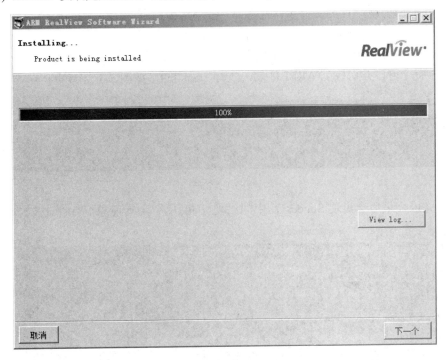

图 1.3-22

(15) 单击"Finish"按钮，如图 1.3-23 所示。

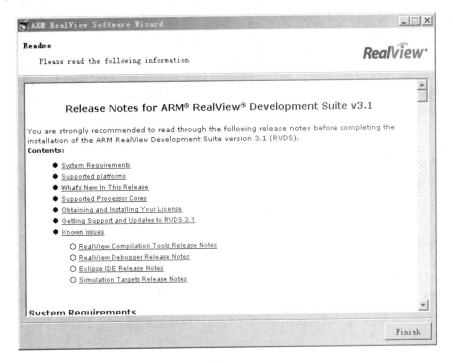

图 1.3-23

到这里 RVCT 已经安装完成了，在第(4)步刚启动 RVCT 时出现的那个 DOS 界面，有的电脑会在后面出现一行英文提示，意思为"按空格键退出"，此时可以单击空格键退出这个 DOS 界面；但有的电脑不会出现这个提示，此时可以强制退出。

(16) 将 RVCT 3.1\rvds4cr 目录下的所有文件都拷贝到 C:\Program Files\ARM 目录下，进入 DOS 命令窗口并切换到 ARM 目录，执行 crack.bat，如图 1.3-24 所示。

图 1.3-24

执行完成，最后会出现 Success，中间出现 Fail 属于正常现象，如图 1.3-25 所示。

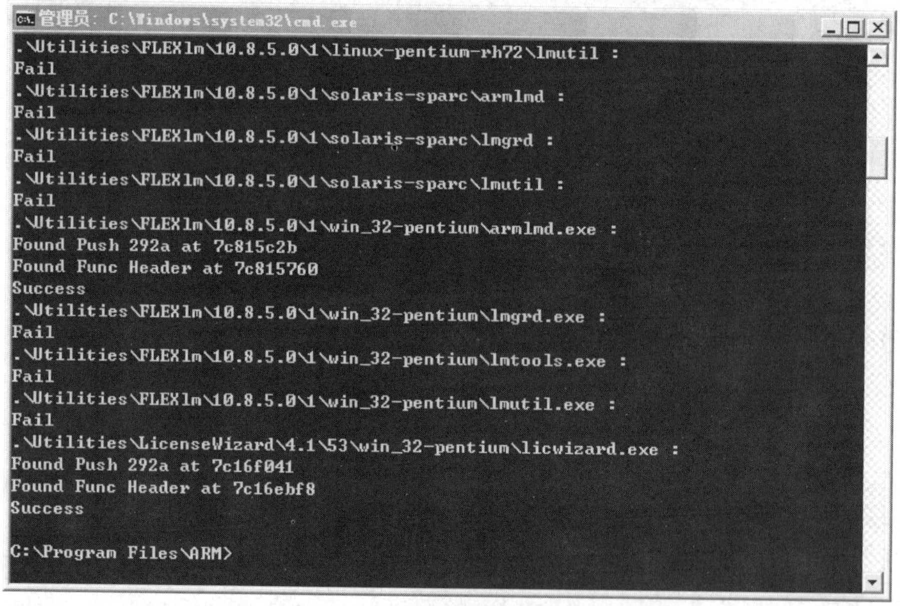

图 1.3-25

把 RVCT 3.1\armar 目录下的 armar.exe 文件拷贝到 C:\Program Files\ARM\RVCT\Programs\3.1\569\win_32-pentium 目录下，替换原来的文件。如果没有替换，则在编译完之后会出现链接错误。

1.3.2 安装 ActivePerl

ActivePerl 是一个 Perl 脚本解释器，它的安装比较简单，笔者以 ActivePerl-5.8.8 版本为例，读者可以使用其他更高的版本。双击运行 ActivePerl-5.8.8.87.2-MSWin32-x86-280952.msi 文件，然后依次单击"下一步"按钮，全部按照默认配置安装即可，直至安装结束。

安装完成之后，就可以进入 CMD 界面，输入"perl –v"，然后按下回车(Enter)键，如果能够查看到 ActivePerl 的版本号，则说明安装成功，如图 1.3-26 所示。

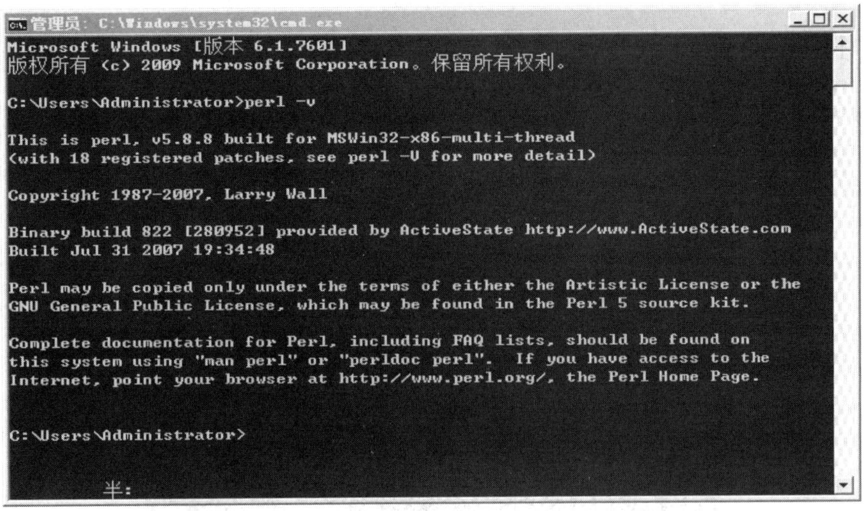

第一章 开 发 准 备

图 1.3-26

1.3.3 安装 Office

在编译 MTK 代码的过程中，需要用到 Office 办公软件中的 Excel 表格，因此还需要安装 Office 办公软件。因为 MTK 中的 Excel 表格后缀名为"xml"，所以建议安装 Office 2003，这个程序的安装本书就不作演示了。

1.3.4 安装 USB 驱动

分别执行 ComPortDriver\InstallDriver.exe 和 ModemPortDriver\ModemInstaller.exe 文件，弹出如图 1.3-27、图 1.3-28 所示的提示框，特别注意提示框中显示的操作系统是否与本机系统对应，否则安装不成功。

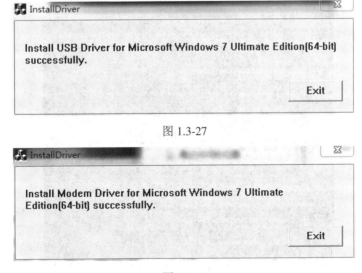

图 1.3-27

图 1.3-28

至此，MTK 编译环境已经安装完成。

1.3.5 环境检测

找到 MTK 的工程目录,在 tool 目录下有一个 chk_env.exe 文件,在 DOS 命令下执行,如图 1.3-29 所示。

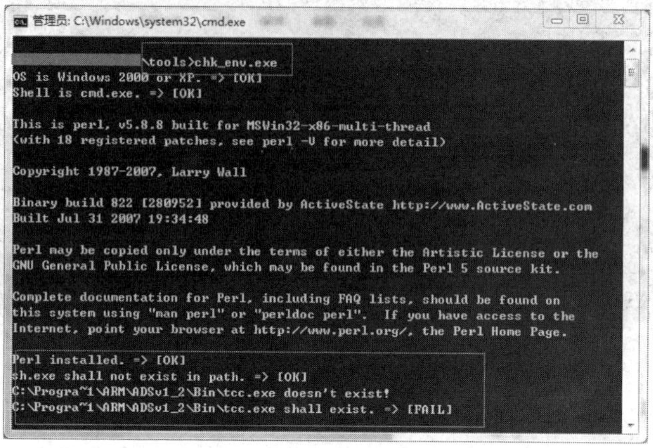

图 1.3-29

因我们并没有安装 ADS,所有运行结果中会出现 [FAIL],这个结果说明环境已经安装成功,但并不代表环境一定是可以使用的。下面再作进一步分析。

1.3.6 编译环境错误分析

在安装 RVCT 过程中,如果在执行 setup.exe 时出现类似于图 1.3-30 所示的界面,可能的原因如下:

(1) 电脑 CPU 为 AMD 速龙系列,无法安装 RVCT,此种情况,只能更换电脑了。

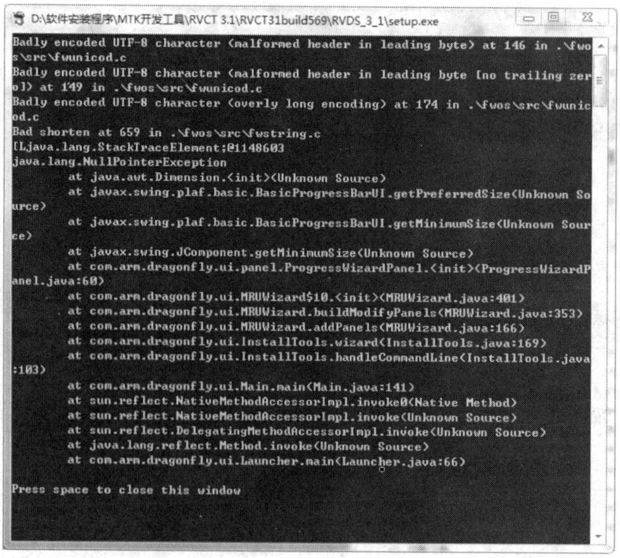

图 1.3-30

(2) 由杀毒软件引起，或没有把电脑主题设置为"Windows 经典"，可重新执行安装步骤中的(1)、(2)。如果依旧无法安装，则尝试重装系统，不要更新任何系统补丁及漏洞，在一份干净的系统下安装。

(3) 如果在编译中出现图 1.3-31 所示的界面，但进入 build\mmi_check.log 文件中却提示"系统找不到指定的路径"(如图 1.3-32 所示)，而执行 tools\chk_env.exe 又显示正常，则可能是操作系统引起的，应重装系统，在一份干净的系统中安装 RCVT。部分网上下载的操作系统安装后也会出现这个错误，这是由于操作系统破解或者漏洞补丁引起的，可以尝试更换其他的操作系统安装包，不要使用 WIN7 家庭版。

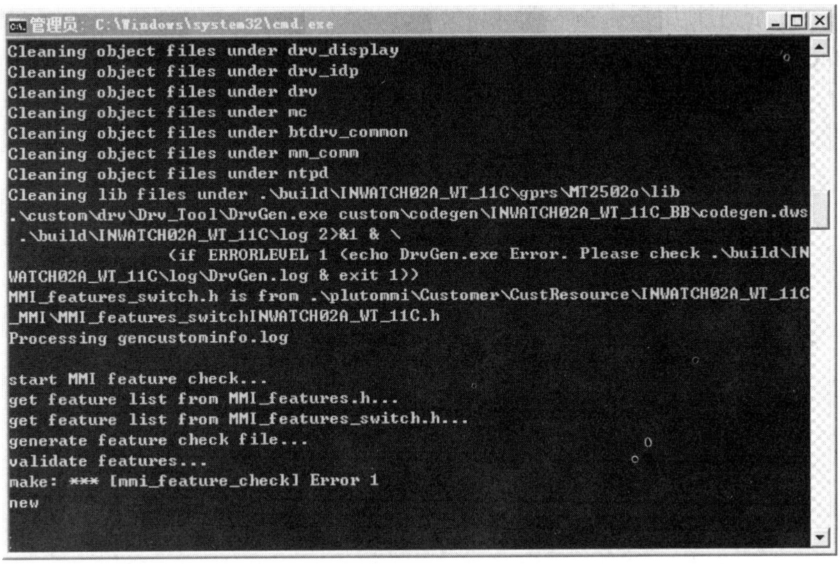

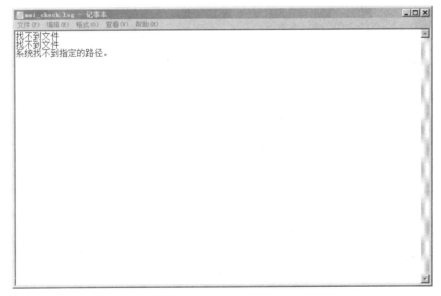

图 1.3-31

图 1.3-32

1.3.7 其他工具软件

在 MTK 开发过程中，除了编译环境之外，还需要用到其他一些工具软件，但这些工具软件有的并不是 MTK 专用的，有的不需要安装，故不再详细介绍安装过程，在后面的开发过程中会介绍其具体用法。这些工具软件主要有：

(1) Source Insight：一款强大的代码编辑工具，在 MTK 开发过程中，大多数代码都是在这个软件中编辑的。

(2) Microsoft Visual Studio 2008：微软开发的 C/C++ 编程工具，MTK 的代码基本都是 C 语言编写的，此工具用于模拟器调试。

(3) Beyond Compare：用于比较代码文件的差异。

(4) FlashTool：烧录工具，将编译代码生成的二进制文件烧录到手机中。

(5) Catcher：用于代码调试，可以打印程序执行的 Log，发送 AT 指令等。

(6) SN_Writer：用于给设备写 IMEI 号码、SN 序列号等。

第二章 开发基础

2.1 编译指令

在开始进入编码之前，先介绍常用的代码编译指令(在接下来的编码过程中，会经常用到)。对于刚接触 MTK 开发的人，可以把这一节的内容当做词典来用，不记得就查一查，没必要死记。在后面的开发中，我们会提示采用什么样的指令编译代码。

编译指令通用格式：

    make [-debug] project project_attr command

指令格式说明：

(1) make：一个批处理的文件名称。在工程源码的根目录下，有一个 make.bat 的文件，在 make 后面的字段实际上都是给批处理传的参数，这个名称一般不允许更改，我们把它当做一个固定字段使用。

(2) -debug：这个参数用得不多，在以前的老版本上编译模拟器的时候需要加上，现在加不加都没关系。

(3) project：我们要编译的项目名称。在工程源码的根目录下，有一个 make 文件夹，在这个文件夹里面有一个 Verno_XXXXXX.bld 文件，比如 Verno_HEXING60A_11B.bld (MT6260A)、Verno_UMEOX60M_11B.bld(MT6260M)。这个文件是项目的版本控制文件，在 Verno_ 后面的 HEXING60A_11B 就是项目名称。在 make 文件夹下面，有可能存在多个这样的文件，那就说明在这个工程源码中，包含多个项目。

(4) project_attr：项目的属性，常用取值有 gprs、gsm、none 等，其中 gprs 代表项目支持 SIM 卡，可以打电话、发短信，还能上网；gsm 表示该项目也支持 SIM 卡，可以打电话，可以发短信，但无法上网，在这样的项目中，我们无法进行 Socket 编程；none 表示项目不支持 SIM 卡。那么如何确定项目的属性呢？在上一条 Project 的介绍中，我们提到了 make 目录下的项目版本控制文件——Verno_HEXING60A_11B.bld，与之对应的还有一个项目配置文件——HEXING60A_11B_XXXX.mak，比如 HEXING60A_11B_GPRS.mak，那么这个项目的属性就是 GPRS。每一个项目在 make 文件夹中都必须包含这两个文件，而且不同的代码项目名称是不一样的，读者拿到任意一份代码，都应该能够找到项目名称。

(5) command：我们要用到的编译命令，采用什么样的方式生成二进制 bin 文件。常用指令的取值及说明见表 2.1-1。

表 2.1-1 常用指令取值及说明

指令	说明
new	这个指令耗时最长，它会重新编译所有模块，每一个项目第一次编译都要使用这个指令，在修改了项目配置或者新增文件时都要使用这个指令重新编译整个项目
Remake(可简写为 r)	这个指令只是简单地重新编译链接有改动的部分，它不检查依赖关系，不扫描资源，只扫描代码的改变，有改变的重编，资源和无改变的代码不编。如果 r 后面没有其他字段，则重新编译所有模块；如果有其他模块字段，则重新编译单个模块或多个模块。比如：r mmi_app 则只重新编译 mmi_app 模块，r mmi_app mmi_framework 则只重新编译两个模块。至于这些模块名称从哪里查找，在后面会有介绍
Updata(可以简写为 u)	这个指令会扫描工程中文件和库的依赖关系，若依赖关系有变化会建立新的依赖关系，随后根据新的依赖关系重新编译链接有改动的部分。它跟 remake 的用法类似，可以更新所有模块，也可以更新指定的模块，比如 u mmi_app, u mmi_app mmi_framework。但是 r mmiresource 不会重新生成资源，u mmiresource 会重新生成资源，相当于 resgen 和 r mmiresource 的合并
Resgen	编译资源。只要修改了.res 文件，或.res 文件中引用到的资源，都要用这个指令编译
Gen_modis	生成 Visual Studio 2008(以下简称 VS2008)模拟器工程文件，可在模拟器上调试 mmi 部分的代码，十分高效快捷
codegen_modis	编译模拟器下的代码文件，不过这个指令的功能可以在 VS2008 中完成，而且在 WIN7 系统下编译无法通过(XP 系统下可以编译通过)，所以我们基本上不用它
new_modis	生成模拟器，实际上就是先执行 Gen_modis，再执行 codegen_modis，在 XP 系统上可以执行通过，但在 WIN7 系统下执行到 codegen_modis 时就会报错，所以我们在 WIN7 系统下也不怎么用它了

编译指令举例：
　　make　HEXING60A_11B GPRS　new
　　make　HEXING60A_11B GPRS　resgen
　　make　HEXING60A_11B GPRS　r
　　make　HEXING60A_11B GPRS　r　mmiresource
　　make　HEXING60A_11B GPRS　u　mmiresource(等同于 make resgen 和 make r mmiresource)
　　make　HEXING60A_11B GPRS　r　mmi_app mmi_framework
　　make　HEXING60A_11B GPRS　u　mmi_app mmi_framework
　　make　-debug　HEXING60A_11B　GPRS　gen_modis

在后面的编码过程中，会省略 project project_attr 用 make command 的格式提示大家用什么样的命令来编译代码。比如 make new 就等同于 make HEXING60A_11B GPRS new，如果你的代码之前有使用该命令编译过，源码根目录中有 HEXING60A_11B_gprs.log 文件，那直接使用 make new 也是可以编译的，同理直接使用 make r 或者 make r mmi_app 都是可以直接编译的。但是对于新手而言，不推荐使用这种简写的指令，一方面，为了防止忘记完整的指令格式；另一方面，如果一个工程中编译了多个项目，则省略项目名称就不知道会编译哪个项目了。

接下来，我们编译本书要用到的 MTK 工程源码。读者下载源码(MTK_STUDY_CODE.zip)解压后，可以看到一个 main 目录，这个 main 目录里面就是整个 MTK 工程源码。首先我们要进入 DOS 命令窗口并切换到 main 目录。(WIN7 系统)进入 DOS 窗口的方法有多种，一种是单击"开始"在"搜索程序和文件"的提示框内直接输入 CMD，然后按回车键；另一种是进入 main 目录，双击"cmd.exe"可执行文件。第二种方法比第一种方法更方便些，因为它打开 DOS 窗口就直接切换到了 main 目录，但"cmd.exe"文件并不是 MTK 的工程文件，如果把它删除，对整个 MTK 工程没有任何影响，而且大多数 MTK 工程都没有这个文件。所以这里给大家介绍一个更简便的方法，按住"Shift"键，鼠标右键单击 main 文件夹，在弹出的右键菜单中选择"在此处打开命令窗口(W)"，就可以进入 DOS 窗口，并切换到了 main 目录下，如图 2.1-1 所示(这种方法在 windows XP 中不支持)。

在 DOS 窗口中输入 make HEXING60A_11B gprs new 指令(如图 2.1-2 所示)，然后按回车键，就会开始编译整个 MTK 工程。编译完成之后，会出现如图 2.1-3 所示界面，如果没有出现该界面，则表示编译出错，至于编译错误如何排查，在后面的章节中会为大家讲解。

图 2.1-1

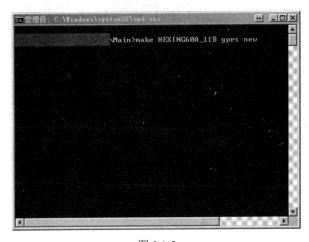

图 2.1-2

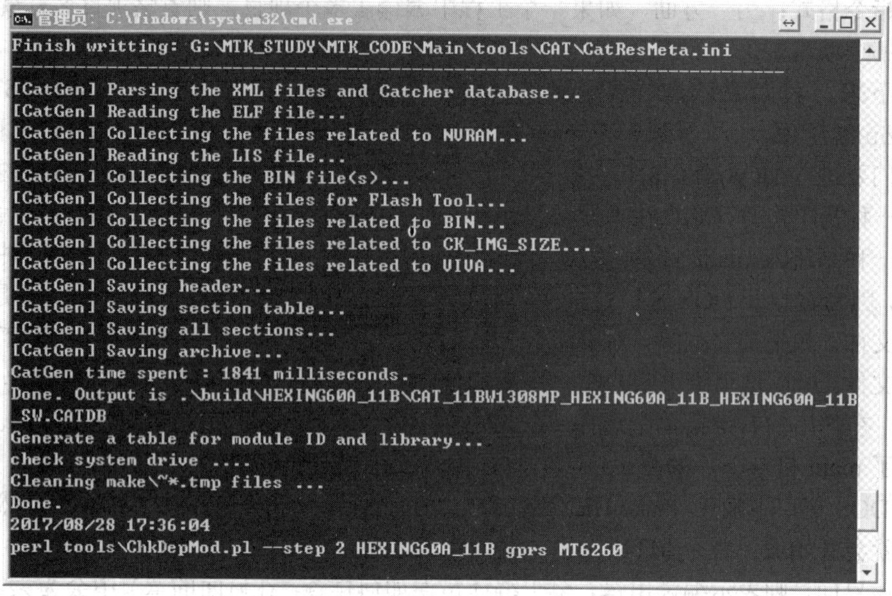

图 2.1-3

上述编译过程完成后，我们就可以把系统下载到 MTK 设备上了，下载方法也放到后面再作讲解。接下来再试一下编译模拟器的指令。直接在打开的 DOS 界面输入 make -debug HEXING60A_11B GPRS gen_modis，然后按下回车键，如图 2.1-4 所示。

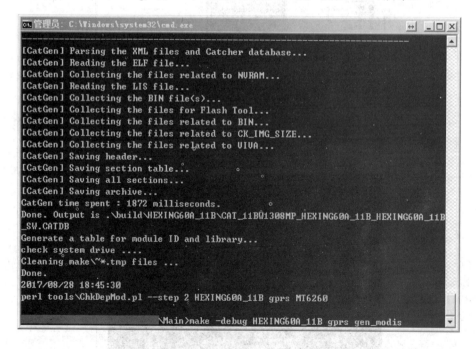

图 2.1-4

编译完成后，会出现如图 2.1-5 所示界面。

图 2.1-5

这个过程通常是不会报错的，除非编译环境没安装成功或者代码不完整。编译完成之后，会在 main\MoDIS_VC9 目录下生成一个 VS2008 的工程文件——MoDIS.sln。我们使用 VS2008 把它打开，然后按下快捷键"F7"编译生成模拟器。如果模拟器不报错，编译完后可以看到如图 2.1-6 所示界面，在底部窗口最后一行字符中能看到"失败 0 个"，如果 VS2008 为英文版的则显示为对应的英文提示。

图 2.1-6

编译完成后，再按下快捷键"Ctrl+F5"直接运行模拟器，或按下快捷键"F5"以调试模式运行模拟器，模拟器显示如图 2.1-7 所示。

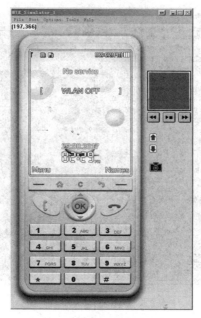

图 2.1-7

2.2 系统框架

一、软件架构

根据本书需要讲解的内容，将 MTK 平台软件架构简要概括，从上到下依次分为 Application 层、Framework 层、Driver 层、Task 层和 Nucleus PLUS 操作系统。

Application
Framework
Driver
Task
NucleusPLUS操作系统

接下来，我们从下到上依次介绍每个层次结构。

1. NucleusPLUS 操作系统

NucleusPLUS 是为实时嵌入式应用而设计的一个抢先式多任务操作系统内核，其 95% 的代码是用 ANSIC 写成的，因此非常便于移植并能够支持大多数类型的处理器。从实现角度来看，NucleusPLUS 是一组 C 函数库，应用程序代码与核心函数库连接在一起，生成一个目标代码，下载到目标板的 RAM 中或直接烧录到目标板的 ROM 中执行。

在典型的目标环境中，NucleusPLUS 核心代码区一般不超过 20 KB 大小。NucleusPLUS 采用了软件组件的方法，每个组件具有单一而明确的目的，通常由几个 C 及汇编语言模块构成，提供清晰的外部接口，对组件的引用就是通过这些接口完成的。除了少数一些特殊情况外，不允许从外部对组件内的全局进行访问。由于采用了软件组件的方法，NucleusPLUS 各个组件非常易于替换和复用。NucleusPLUS 的组件包括任务控制、内存管理、任务间通信、任务的同步与互斥、中断管理、定时器及 I/O 驱动等。

2．Task 层

Task 层负责系统的任务调度，实际上就是一个死循环。MTK 平台中几乎所有的系统消息，最终都是在这一层调用的。开发者可以自己新建 Task，但在实际项目开发中，很少涉及这一层的修改。

3．Driver 层

Driver 层为驱动层，所有的硬件驱动都在这一层调用执行，包括摄像头驱动、LCD 驱动、蓝牙驱动、按键配置等。这一层在开发中是重点，任何驱动调试以及外设添加都要在这一层修改或添加驱动代码。

4．Framework 层

Framework 层主要封装了一些功能接口，为 Application 开发提供支持，包括 gui、gdi、mdi、filemanger、timer 等。在实际开发过程中，使用到的函数接口，大多数来自于这一层。另外，也可以在这一层封装或修改功能接口，比如定制自己的 category screen 函数，实现特定的风格。

5．Application 层

Application 层是应用层，主要负责跟用户交互，与 Framework 层统称为 MMI(The Man-Machine Interface 人机界面)。系统中自带的音乐播放器、通讯录、短信等具体功能都属于一个 Application。在 MTK 平台软件开发中，大部分的工作都集中在这一层。

二、文件结构

如图 2.2-1 所示，这是一个 MTK 系统的完整目录，所示的 MTK 系统版本为 MT6260A，本书所有的示例代码以及后面的实例项目，都是基于 MT6260A 系统平台开发的。

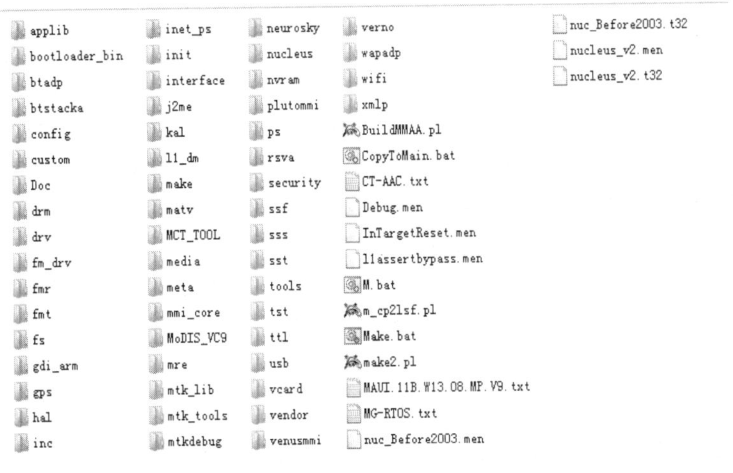

图 2.2-1

所有MTK功能机平台的系统代码文件结构都与图2.2-1所示大同小异,这是没有编译过的源码结构,编译之后还会多出来一些文件(夹),如图2.2-2所示。

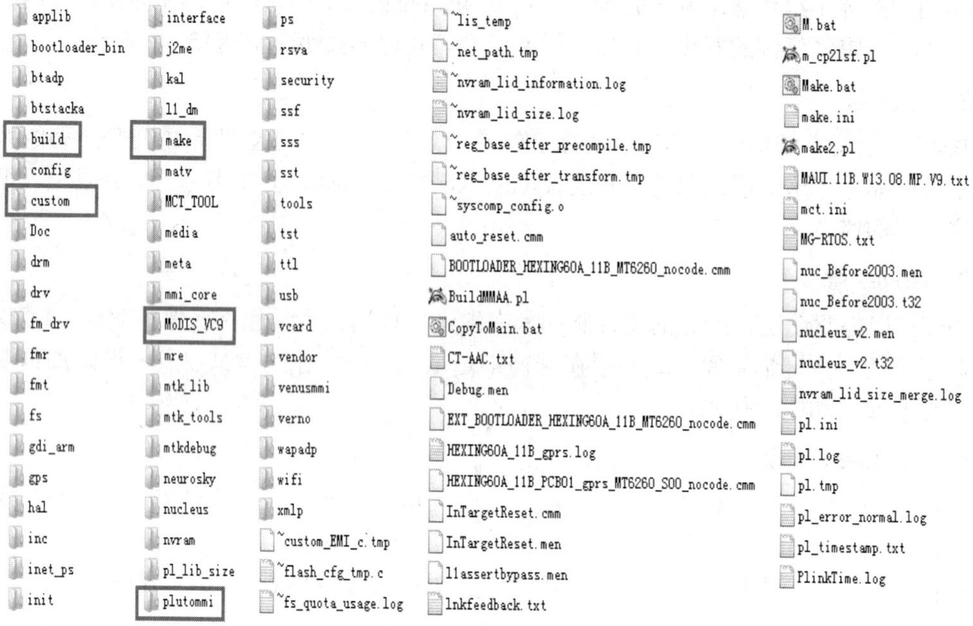

图2.2-2

虽然看上去文件(夹)很多,有点眼花缭乱,但我们需要关注的文件目录并不多,主要有 build、custom、make、MoDIS_VC9、plutommi,在后面的开发过程中,会经常与这几个文件夹打交道,下面分别为大家简单介绍这几个目录。

(1) build:编译生成的目录。在上一章节中,我们编译了HEXING60A_11B项目,那么在 build 目录下会有一个与项目名称一模一样的文件夹——HEXING60A_11B,该文件夹主要存放编译完成之后的烧录文件——HEXING60A_11B_PCB01_gprs_MT6260_S00.HEXING60A_11B_SW.bin 以及编译过程中生成的 log 文件,如果编译出错,一般都是在这个目录下的 log 文件夹中查看对应的".log"文件。

(2) custom:驱动相关的底层代码文件都在这个目录下,新增复合数据的 NVRAM 也会涉及这个目录。

(3) make:在前面讲解编译指令的时候,有提到过这个目录。这个目录包括子目录下的文件,后缀名大多数都是".mak",这些".mak"文件是整个系统的 Makefile 文件,它们对整个系统所有代码的编译进行管理,每个子目录下的".mak"文件名都代表一个模块,可以使用"make r 模块名"(比如:make r custom)单独进行编译。当然,我们还可以新增自己的功能模块,也可以单独编译自己的代码。项目配置文件(HEXING60M_11B_GPRS.mak)也包含在这个目录下,通过修改这个文件,可以定义我们的MTK产品包含哪些功能,比如是否支持 camera(照相机)、audio record(录音)等。还有一个就是版本号(Verno_HEXING60M_11B.bld)文件也在此目录下。

(4) MoDIS_VC9:使用 make gen_modis 编译生成的模拟器 VS2008 工程文件,以及VS2008 中编译生成的 obj 文件和模拟器可执行文件都包含在这个目录下。这个目录下的文件我们基本上不会修改，但如果新增一个屏幕分辨率，在模拟器下调试也要适配对应的分辨率，此时就要在 MoDIS_VC9\MoDIS\Skin 中添加对应的模拟器 UI 尺寸。另外 MTK 设备中的存储功能，在 MoDIS_VC9\WIN32FS 下可以模拟，比如 DRIVE_C 是系统盘，编程中对应的盘符为"C"盘，这个目录在运行模拟器的时候会自动生成，如果修改了 NVRAM 的代码，一般需要删除这个目录，相当于真机设备下载系统时的 format 操作；DRIVE_D 目录是系统分配的磁盘空间，编程中对应的盘符为"D"盘，这个空间在真机设备中不一定存在，它取决于 Flash 的大小，如果 Flash 够大，可以在 custom_MemoryDevice.h 文件中分配，如果分配成功，在没插入 T 卡的情况下，用 USB 连接电脑会显示可移动磁盘；DRIVE_E 目录就是模拟 T 卡的磁盘空间，SIM_CARD 是模拟 SIM 卡。以上四个文件夹只有 DRIVE_C 可以删除，DRIVE_D 和 DRIVE_E 不可删除，但里面的内容可以删除。SIM_CARD 目录禁止删除，否则模拟器会运行不了。

(5) plutommi：这个目录是本书的重点，它里面几乎包含了 MMI(Man Machine Interface 即人机界面)层的所有源代码文件和资源文件。

2.3 新增功能模块

MTK 系统的代码大部分都是用 C 语言编写的，少部分使用 C++，但我们很少在上面编写 C++ 代码。C 语言是面向过程的编程语言，程序的功能都是通过一个或多个函数实现的，各个函数之间应符合"高内聚，低耦合"的软件设计标准，这一标准对于面向过程的编程语言尤其重要。一个好的 C 程序不仅要注重代码的执行效率，还要注重代码的可读性及可维护性。MTK 的代码在这方面做得很好，它的所有功能都是模块化管理，几乎每个功能都有对应的功能宏开关，通过配置宏开关，就可以实现代码的选择性编译，从而定制我们的 MTK 产品自带哪些功能。

对于好的东西，我们应该学习并将其发扬光大。如果从事 MTK 功能手机的研发，那我们大多数的工作都是对 MTK 平台现有的功能进行客制化修改，很少需要自己重新开发一个功能。但从事 MTK 平台物联网或智能硬件产品的研发，大多数功能都需要自己从零开始编码实现。对于新开发的功能，我们也用一个功能宏控制起来，这样既方便移植，也方便代码管理。然而一个 MTK 产品并不只有一个功能，我们需要开发多个功能，那么把自己开发的所有功能独立出来管理，就形成一个模块。这个模块我们可以单独对它进行编译，也能在一定程度上节省编译时间。

在还没有正式进入编程之前，需要对 MTK 的框架进行扩展，对于刚接触 MTK 的人来说有点晦涩难懂。但没有关系，你只需要记住如何新增一个模块就行了，没必要深究为什么这么做。之所以把这章放到正式编码之前，也是符合软件设计的流程：先搭框架，再写代码。具体步骤如下：

(1) 首先，给 MTK 项目建一个 Source Insight(请读者自己安装，使用方法请查阅附录 A 工程，打开 Source Insight，点击菜单 Project→New Project，在弹出的对话框中输入项目名称(MTK_STUDY_CODE)和项目文件保存的路径(笔者存放在 G:\MTK_STUDY\MTK_STUDY_CODE 中)，如图 2.3-1 所示。

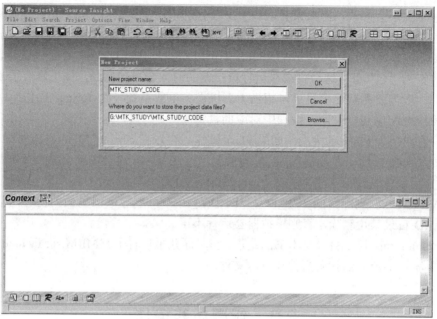

图 2.3-1

(2) 单击"OK"按钮,在弹出的对话框中再单击"OK"按钮,在"Add and Remove Project Files"对话框中,加载项目文件,选择"Mtk_study_code",再选择右侧的"Main",然后单击"Add Tree",如图 2.3-2 所示。

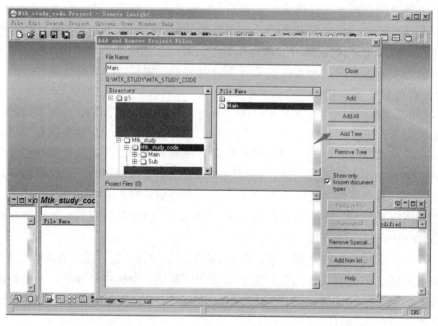

图 2.3-2

(3) 加载完后,会弹出一个如图 2.3-3 所示对话框,提示总共加载了多少个文件,图示为 20095 个文件。单击"确定"按钮,然后在"Add and Remove Project Files"窗口中单击"Close"按钮。到这里 Source Insight 工程就建好了。

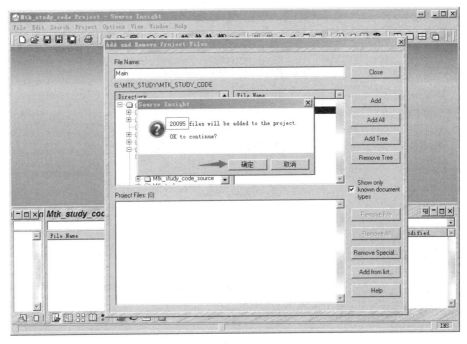

图 2.3-3

我们在 Source Insight 里面随便打开一个 .c 文件，比如 mainmenu.c，如图 2.3-4 所示。

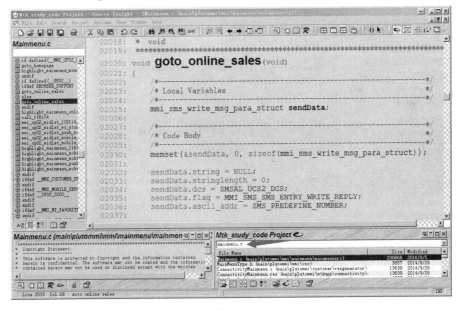

图 2.3-4

我们发现 goto_online_sales 函数中有些变量和数据类型是灰色的，而且无法快速找到它们定义的地方(按住 Ctrl 键，再用鼠标点击变量或数据类型或函数名称能快速查找到定义的地方)，这是因为刚建立的 Source Insight 工程还没有同步所有工程文件，需要 Rebuild 一下。点击 Project 菜单，选择 Rebuild Project，在弹出的对话框中选择"Re-Parse all source files"，单击"OK"按钮，如图 2.3-5 所示。

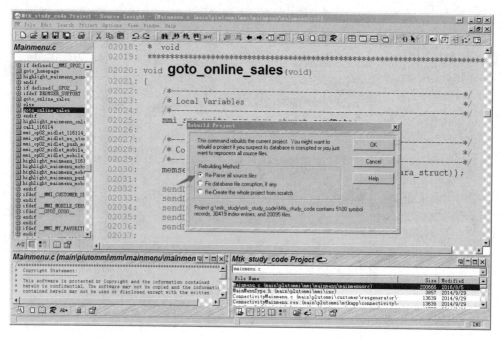

图 2.3-5

同步完成后，再看看 mainmenu.c 文件，会发现变量或数据类型的颜色变了，而且也能快速找到定义的地方，如图 2.3-6 所示。

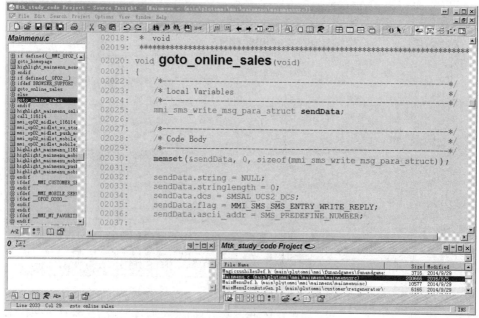

图 2.3-6

接下来我们就新增一个自己的模块，命名为：mystudy，在后面的编程中，我们所有的功能代码都只在自己的模块中编写。

在 make\HEXING60A_11B_GPRS.mak 文件中添加整个模块的宏开关 MYSTUDY_

APPLICATION，宏定义命名都用大写。代码如下(代码清单 2.3-1)：

--代码清单 2.3-1--

```
/* 省略部分代码*/
PHB_NAME_LENGTH = 80


#------------------------------------------------------##
#---------- mystudy add begin

MYSTUDY_APPLICATION = TRUE              # 定义模块宏

#---------- custom add end
#------------------------------------------------------##


# ****************************************************************
# Release Setting Section
# ****************************************************************

/* 省略部分代码*/
```

代码添加的位置，可以是任意位置，但必须放在"include make\MODEM.mak"语句的前面，这里我们把它加在"Release Setting Section"注释语句前面。

在 make\option.mak 文件的末尾处添加如下代码(代码清单 2.3-2)：

--代码清单 2.3-2--

```
/* 省略部分代码*/
ifeq ($(strip $(CUSTOM_RELEASE)), TRUE)
   ifeq ($(strip $(MTK_SUBSIDIARY)), TRUE)
      RELEASE_LEVEL = SUB_RELEASE
   endif
endif


#-----------------__MYSTUDY_APPLICATION__----------------##
#------------------------ mystudy add begin


# ***************************************************
# 功能宏
# ***************************************************
ifdef MYSTUDY_APPLICATION
    ifeq ($(strip $(MYSTUDY_APPLICATION)), TRUE)
        COMPLIST     += mystudyapp
```

```
            CUS_REL_BASE_COMP += make\mystudyapp
            COM_DEFS        += __MYSTUDY_APPLICATION__
        endif
    endif

    #---------- mystudy add end
    #------------------------------------------------------##
    -----------------------------------------------------------------------------------
```

注意：

(1) ifdef MYSTUDY_APPLICATION 和 endif 是成对出现的，如果定义了 MYSTUDY_APPLICATION 宏，就会执行 ifdef 和 endif 包含的代码。

(2) ifeq ($(strip $(MYSTUDY_APPLICATION)), TRUE)和 endif 也是成对出现的，意思是如果 MYSTUDY_APPLICATION 宏的值等于 TRUE，则执行 ifeq 和 endif 包含的代码。如果不等于 TRUE(等于 FALSE)则不执行宏包含的代码，这样就实现了宏开关对模块的控制。

(3) COMPLIST   += mystudyapp
    CUS_REL_BASE_COMP += make\mystudyapp
    COM_DEFS   += __MYSTUDY_APPLICATION__

这三行代码，前两行就是添加模块 mystudyapp，让该模块参与到编译中，最后一行是定义模块的功能宏，没有这条语句也不会影响模块的添加，但为了方便代码控制，最好还是加上，后面模块相关的所有代码都可以用这个宏包含。此处需特别注意，模块名"mystudyapp"必须全部是小写字符，如果包含大写字符，则编译报错。

(4) 在(3)中有一句代码"CUS_REL_BASE_COMP += make\mystudyapp"，这里指定了 mystudyapp 模块的 Makefile 路径在 make\mystudyapp 目录中，所以我们需在 make 目录下新建一个文件夹"MyStudyApp"(文件名不区分大小写)，在"MyStudyApp"文件夹里面再新建一个"MyStudyApp.mak"文件，这个文件就是 mystudyapp 模块的 Makefile 文件，它指定了模块中的头文件路径以及需要编译的源文件。那么这个 Makefile 如何写呢？因为 Makefile 文件的编写并不是本章要介绍的内容，这里教给大家一个取巧的方法：把 make\plutommi\mmi_app\mmi_app.mak 文件中的内容全部复制到 MyStudyApp.mak 文件中，然后把 SRC_LIST 指定的源文件(.c)全部删除，只留下 INC_DIR 指定的头文件目录。最终 MyStudyApp.mak 文件内容如下(代码清单 2.3-3)：

```
-----------------------------------------代码清单 2.3-3-----------------------------------------
# Define include path lists to INC_DIR
INC_DIR = applib\inet\engine\include \
          applib\mem\include \
          applib\misc\include \
          drv\include \
          fmt\include \
          fs\fat\include \
```

```
interface\hwdrv \
interface\wifi \
$(strip $(PS_FOLDER))\interfaces\local_inc \
plutommi\customer\customerinc \
plutommi\customer\customize \
plutommi\customer\custresource \
plutommi\mmi\asyncevents\asynceventsinc \
plutommi\mmi\inc \
plutommi\Framework\Interface \
plutommi\mmi\inc\menuid \
plutommi\Framework\CommonFiles\commoninc \
plutommi\Framework\EventHandling\eventsinc \
plutommi\Framework\History\historyinc \
plutommi\Framework\MemManager\memmanagerinc \
plutommi\Framework\NVRAMManager\nvrammanagerinc \
plutommi\Framework\Tasks\tasksinc \
plutommi\Framework\ThemeManager\thememanagerinc \
plutommi\Framework\GUI\gui_inc \
plutommi\Framework\GUI\oem_inc \
plutommi\mmi\miscframework\miscframeworkinc \
plutommi\Framework\InputMethod\Inc \
plutommi\Framework\GDI\gdiinc \
plutommi\service\mdi\mdiinc \
plutommi\Service\ImgEdtSrv \
plutommi\mtkapp\mtkappinc \
vendor\inputmethod\cstar\adaptation\include \
vendor\inputmethod\t9\adaptation\include \
venusmmi\app\pluto \
btstacka\inc \
drm\include \
gps\inc \
irda\inc \
j2me\interface \
j2me\jal\include \
j2me\jal\include \
media\image\include \
media\rtp\inc \
media\stream\include \
media\mtv\src \
media\mtv\include \
```

```
usb\include \
verno \
wapadp\include \
xmlp\include \
vcard\include \
plutommi\CUI\Inc \
plutommi\CUI\InlineCui \
plutommi\mmi\athandler\athandlerinc \
plutommi\mmi\audio\audioinc \
plutommi\Framework\BIDI\bidiinc \
plutommi\mmi\browserapp\browser\browserinc \
plutommi\mmi\browserapp\pushinbox\pushinboxinc \
plutommi\MMI\CertificateManager\CertificateManagerMMIInc \
security\certman\include \
plutommi\Framework\CommonScreens\commonscreensinc \
plutommi\MMI\ConnectManagement\ConnectManagementInc \
plutommi\mmi\cphs\cphsinc \
plutommi\mmi\customerservice\customerserviceinc \
plutommi\mmi\dataaccount\dataaccountinc \
plutommi\mmi\datetime\datetimeinc \
plutommi\Framework\DebugLevels\debuglevelinc \
plutommi\mmi\dictionary\dictinc \
plutommi\mmi\ebookreader\ebookinc \
plutommi\mmi\emailapp\emailappinc \
plutommi\service\ProvisioningSrv \
plutommi\mmi\extra\extrainc \
plutommi\mmi\funandgames\funandgamesinc \
plutommi\mmi\gpio\gpioinc \
plutommi\mmi\gsmcallapplication\commonfiles\commoninc \
plutommi\mmi\gsmcallapplication\incomingcallmanagement\incominginc \
plutommi\mmi\gsmcallapplication\outgoingcallmanagement\outgoinginc \
plutommi\mmi\help\helpinc \
plutommi\mmi\idlescreen\idlescreeninc \
plutommi\mmi\idlescreen\idlescreeninc\demoappinc \
plutommi\MMI\Factory\FactoryInc \
plutommi\MMI\Idle\IdleInc \
plutommi\MMI\Dialer\DialerInc \
plutommi\MMI\ScrLocker\ScrLockerInc \
plutommi\MMI\ScrSaver\ScrSaverInc \
plutommi\mmi\imps\impsinc \
```

plutommi\Framework\IndicLanguages\indiclanguagesinc \
plutommi\Framework\Languages\thai\thaiinc \
plutommi\mmi\mainmenu\mainmenuinc \
plutommi\mmi\messages \
plutommi\CUI\SsoCui\SsoCuiInc \
plutommi\MMI\SSOAPP\SSOAPPInc \
plutommi\mmi\messages\messagesinc \
plutommi\mmi\mobileservice\mobileserviceinc \
plutommi\mmi\nitzhandler\nitzinc \
plutommi\mmi\organizer\Organizerinc \
plutommi\mmi\Organizer\IndianCalendar\IndianCalendarInc \
plutommi\mmi\Organizer\Reminder \
plutommi\mmi\phonebook\phonebookinc \
plutommi\mmi\phonebook\core \
plutommi\mmi\profiles\profilesinc \
plutommi\mmi\resource\inc \
plutommi\mmi\sat\satinc \
plutommi\mmi\setting\settinginc \
plutommi\mmi\SecuritySetting\SecSetInc \
plutommi\mmi\smartmessage\smartmessageinc \
plutommi\mmi\ucm\ucminc \
plutommi\mmi\BT_UCM\UcmBTInc\
plutommi\mmi\ctm\ctminc \
plutommi\service\Inc \
plutommi\service\UmSrv \
plutommi\mmi\unifiedmessage\unifiedmessageinc \
plutommi\Service\UcSrv \
plutommi\mmi\unifiedcomposer\unifiedcomposerinc \
venusmmi\app\pluto_variation\CubeApp \
plutommi\mmi\CubeApp\CubeAppInc \
plutommi\mmi\unifiedmms\mmsapp\mmsappinc \
plutommi\Service\UmmsSrv \
plutommi\mmi\voip\voipinc \
plutommi\mmi\Organizer\HijriCalendar\HijriCalendarInc \
plutommi\mtkapp\abrepeater\abrepeaterinc \
plutommi\MtkApp\AGPSLog\AGPSLogInc \
plutommi\mtkapp\audioplayer\audioplayerinc \
plutommi\mtkapp\MediaPlayer\MediaPlayerInc \
plutommi\mtkapp\avatar\avatarinc \
plutommi\mtkapp\barcodereader\barcodereaderinc \

```
plutommi\mtkapp\bgsound\bgsoundinc \
plutommi\mtkapp\camera\camerainc \
plutommi\MtkApp\ImageView\ImageViewInc \
plutommi\mtkapp\camcorder\camcorderinc \
plutommi\mtkapp\centralconfigagent\centralconfigagentinc \
plutommi\mtkapp\connectivity\connectivityinc \
plutommi\mtkapp\connectivity\connectivityinc\btcommon \
plutommi\mtkapp\connectivity\connectivityinc\btmtk \
plutommi\Framework\CSBrowser\csbrowserinc \
plutommi\mtkapp\DCD\DCDInc \
plutommi\mtkapp\dlagent\dlagentinc \
plutommi\mtkapp\dmuiapp\dmuiappinc \
plutommi\mtkapp\EngineerMode\EngineerModeApp\EngineerModeAppInc \
plutommi\mtkapp\EngineerMode\EngineerModeEngine\EngineerModeEngineInc \
plutommi\mtkapp\FactoryMode\FactoryModeInc \
custom\common\hal \
plutommi\mtkapp\filemgr\filemgrinc \
plutommi\mtkapp\fmradio\fmradioinc \
plutommi\mtkapp\fmschedulerec\fmschedulerecinc \
plutommi\mtkapp\GPS\GPSInc \
plutommi\mtkapp\javaagency\javaagencyinc \
plutommi\mtkapp\mmiapi\include \
plutommi\mtkapp\mobiletvplayer\mobiletvplayerinc \
plutommi\MtkApp\MobileTVPlayer\MtvSMSInc \
plutommi\mtkapp\dtvplayer\dtvplayerinc \
plutommi\mtkapp\myfavorite\myfavoriteinc \
plutommi\mtkapp\photoeditor\photoeditorinc \
plutommi\mtkapp\pictbridge\pictbridgeinc \
plutommi\mtkapp\rightsmgr\rightsmgrinc \
plutommi\MtkApp\SoftwareTracer\SoftwareTracerInc \
plutommi\mtkapp\soundrecorder\soundrecorderinc \
plutommi\mtkapp\sndrec\sndrecinc \
plutommi\mtkapp\answermachine\answermachineinc \
plutommi\mtkapp\swflash\swflashinc \
plutommi\mtkapp\syncml\syncmlinc \
plutommi\mtkapp\video\videoinc \
plutommi\mtkapp\vobjects\vobjectinc \
plutommi\mtkapp\vrsd\vrsdinc \
plutommi\mtkapp\vrsi\vrsiinc \
plutommi\mtkapp\searchweb\searchwebinc \
```

```
plutommi\vendorapp\devapp\devappinc \
vendor\dict\gv\adaptation\inc \
vendor\dict\motech\adaptation\inc \
vendor\dict\trilogy\adaptation\inc \
vendor\game_engine\brogent\adaptation \
vendor\game_engine\brogent\game \
vendor\game_engine\intergrafx\adaptation \
vendor\game_engine\intergrafx\game \
vendor\swflash\neomtel\adaptation\inc \
vendor\wap\obigo_q05a\adaptation\integration\owl\include \
vendor\wap\obigo_q05a\adaptation\modules\bam\include \
vendor\wap\obigo_Q03C\v1_official\modules\bra\config \
vendor\wap\obigo_Q03C\v1_official\modules\bra\refsrc \
vendor\wap\obigo_Q03C\v1_official\modules\bra\source \
vendor\wap\obigo_Q03C\v1_official\modules\mea\intgr \
vendor\wap\obigo_Q03C\v1_official\msf\msf_lib\config \
vendor\wap\obigo_Q03C\v1_official\msf\msf_lib\export \
vendor\wap\obigo_Q03C\v1_official\msf\msf_lib\intgr \
vendor\wap\obigo_Q03C\adaptation\modules\mma\include \
vendor\wap\obigo_Q03C\adaptation\msf_ui\include \
vendor\mobilevideo\adaption\MobileVideoInc\include_CM \
vendor\mobilevideo\adaption\MobileVideoInc\include \
plutommi\mmi\SIMProvAgent\SIMProvAgentInc \
plutommi\mmi\UDX\UDXInc \
vendor\qqim\adaptation\inc \
plutommi\Service\ProvBoxSrv \
plutommi\mmi\ProvisioningInbox\ProvBoxApp\ProvBoxAppInc \
init\include \
$(strip $(PS_FOLDER))\l4\include \
venusmmi\vrt\interface \
venusmmi\framework \
venusmmi\framework\interface \
venusmmi\visual \
venusmmi\visual\interface \
venusmmi\app\pluto_variation\interface \
venusmmi\app\pluto_variation\adapter \
venusmmi\app\pluto_variation\adapter\interface \
venusmmi\app\pluto_variation\adapter\interface\res \
venusmmi\test \
venusmmi\app \
```

```
venusmmi\app\pluto_variation \
venusmmi\app\common\wallpaper \
venusmmi\app\common\ncenter \
venusmmi\app\common\interface\app \
venusmmi\framework\ui_core\mvc \
venusmmi\app\common\data \
plutommi\mtkapp\MREAPP\MREAPPInc \
plutommi\mtkapp\MSpace\MSpaceInc \
plutommi\Service\MRESrv \
plutommi\CUI\ToneSelectorCui \
plutommi\Framework\LangModule\LangModuleInc \
plutommi\MMI\Bootup\BootupInc \
plutommi\MMI\NwInfo\NwInfoInc \
plutommi\MMI\SimCtrl\SimCtrlInc \
plutommi\MMI\Shutdown\ShutdownInc \
plutommi\mmi\q03c_mms_V01_agent\q03c_mms_V01_agentinc \
plutommi\Service\MediaCacheSrv \
plutommi\Service\inc \
plutommi\MtkApp\WgtMgrApp\WgtMgrAppInc \
plutommi\MMI\CallSetting\CallSettingInc \
plutommi\MMI\SupplementaryService\SsInc \
vendor\opera\browser\adaptation\inc \
plutommi\Service\NetSetSrv \
plutommi\Service\ModeSwitchSrv \
plutommi\Service\ProfilesSrv \
plutommi\MMI\ShellApp\ShellAppInc \
venusmmi\app\pluto_variation\ShellApp\base \
venusmmi\app\pluto_variation\ShellApp\panel \
venusmmi\app\pluto_variation\adapter\ShellApp\base \
venusmmi\app\pluto_variation\adapter\ShellApp\panel \
plutommi\mmi\CallLog\CallLogInc \
venusmmi\app\pluto_variation\adapter\ShellApp\panel\HomeScreen \
venusmmi\app\pluto_variation\adapter\ShellApp\panel\MsgViewer\
venusmmi\app\pluto_variation\ShellApp\panel\HomeScreen \
venusmmi\app\pluto_variation\adapter\MultiTouchTest \
vendor\opera\browser\v1_official\opdev\include \
operator\ORANGE\common\venusmmi\pluto_adapter\interface\res \
operator\ORANGE\common\venusmmi\pluto_adapter\HomeScreen \
operator\ORANGE\common\venusmmi\app\HomeScreen \
operator\ORANGE\common\venusmmi \
```

mmi_core\app\SupplementaryService \
plutommi\mmi\SimSpace\SimSpaceInc \
plutommi\CUI\SimSelCui \
plutommi\Framework\InputMethod\Engine\Engine_Inc \
plutommi\Framework\InputMethod\SDK_Layer\SDK_Inc \
plutommi\Framework\InputMethod\UI\UI_Inc \
hal\storage\mc\inc \
plutommi\mtkapp\Lemei\LemeiInc \
plutommi\Service\Inc \
plutommi\Service\UPPSrv \
hal\video\custom \
custom\video \
interface\hal\display\common \
interface\ps\enum \
meta\cct \
plutommi\Service\DmSRsrv \
vendor\widget\google\adaptation \
interface\nfc \
plutommi\CUI\UseDetailCui \
interface\hal\video_codec \
custom\drv\measure\Inc \
custom\common\hal_public \
custom\codegen\$(strip $(BOARD_VER))\
interface\hal\graphics \
interface\hal\fmr \
plutommi\Service\RightsMgr \
plutommi\AppCore\SSC \
venusmmi\app\pluto_variation\LauncherKey\widget\musicplayer \
plutommi\AppCore\Ucm \
plutommi\cui\ImageClipCui\ImageClipCuiInc \
plutommi\Service\TcardSrv \
plutommi\MtkApp\SafeMode\SafeModeInc \
hal\system\dcmgr\inc \
plutommi\Service\SmsSrv \
plutommi\Service\SmsbtmapcSrv \
plutommi\mmi\SmsbtmapcApp\SmsbtmapcAppInc \
plutommi\MMI\BTNotification\BTNotificationInc

ifneq ($(filter __MMI_LAUNCHER_APP_LIST__ , $(strip $(MODULE_DEFS))),)
    INC_DIR += plutommi\MMI\AppList\AppListInc

```
        endif

        ifeq ($(filter __MTK_TARGET__ , $(strip $(MODULE_DEFS))), )
            INC_DIR += MoDIS_VC9\MoDIS    # for w32_utility.h
        endif

        ifneq ($(filter __RF_DESENSE_TEST__, $(strip $(MODULE_DEFS))), )
           INC_DIR += BTMT\rf_desense\$(strip $(PLATFORM))
        endif
        ifneq  ($(filter __SOCIAL_NETWORK_SUPPORT__ , $(strip $(MODULE_DEFS))), )

            INC_DIR += plutommi\mmi\sns\snsinc
            INC_DIR += plutommi\cui\SnsCui
        endif

        ifneq ($(filter __MMI_VPP_UPGRADE__, $(strip $(MODULE_DEFS))), )
            INC_DIR += plutommi\MMI\Upgrade\UpgradeInc
        endif
```

到这里，mystudyapp 模块已经添加成功了，但是 make new 编译的时候依旧无法通过，因为模块里面只有头文件，没有源文件，编译出来的 lib 文件是空的。

因此，我们要在 plutommi 目录下新建 MyStudyApp 文件夹，然后在 MyStudyApp 中新建 MyStudyAppMain 文件夹，接着在该文件夹里面再新建一个 Inc 和 Src 文件夹，最终目录结构如图 2.3-7 所示。

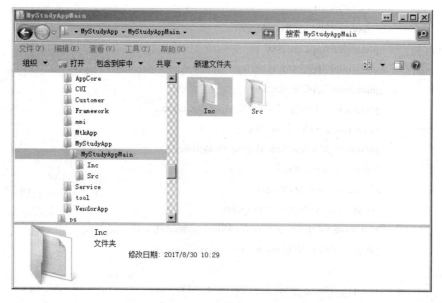

图 2.3-7

在 Inc 中新建头文件 MyStudyAppMain.h，在 Src 中新建源文件 MyStudyAppMain.c。这两个文件暂时都置为空文件，里面不用添加任何内容。

在 MyStudyApp.mak 文件末尾处添加源文件 MyStudyAppMain.c 并指定 MyStudyAppMain.h 头文件的路径，把代码用_MYSTUDY_APPLICATION_宏控制起来，新增代码如下(代码清单 2.3-4)：

--代码清单 2.3-4--
```
/*省略部分代码*/
ifneq ($(filter __MMI_VPP_UPGRADE__, $(strip $(MODULE_DEFS))), )
    INC_DIR += plutommi\MMI\Upgrade\UpgradeInc
endif


ifneq ($(filter __MYSTUDY_APPLICATION__ , $(strip $(MODULE_DEFS))), )
    INC_DIR += plutommi\MyStudyApp\MyStudyAppMain\Inc          #指定头文件路径
    SRC_LIST += plutommi\MyStudyApp\MyStudyAppMain\Src\MyStudyAppMain.c #新增源文件
endif
```

以上步骤执行完之后，需使用 make new 编译，编译结束则在 build\HEXING60A_11B\gprs\MT6260o\能看到 MyStudyApp 文件夹，并包含 MyStudyAppMain.obj 文件，说明 mystudyapp 模块已经参与到了系统编译中，并且编译了源文件 MyStudyAppMain.c。在 build\HEXING60A_11B\log 目录中会生成 MyStudyApp.log 和 mystudyapp_setEnv.log 文件。如果 mystudyapp 模块报错，则在 MyStudyApp.log 文件中查看错误信息。新增模块后，模拟器也需要重新生成，执行 make -debug HEXING60A_11B gprs gen_modis 指令，并用 VS2008 编译，可以在 MoDIS_VC9 文件夹中看到 mystudyapp 目录，说明模拟器里面也编译了 mystudyapp 模块。

新增一个模块其实并不难，总结一下就以下几个步骤：

(1) HEXING60A_11B_GPRS.mak 文件中添加模块宏开关。
(2) option.mak 文件中指定模块名称以及模块 makefile，并定义模块宏。
(3) make 目录下创建一个以模块名命名的文件夹及同名的 Makefile 文件(.mak)。
(4) 新建模块的代码文件。(模块代码文件的位置可以放在工程里的任意位置，但必须在 Makefile 文件中能找到，笔者选择 plutommi 目录)
(5) 把模块的代码文件添加到模块的 Makefile(.mak)文件中。

本章的代码，在下载的源码对应章节目录中，双击"CopyToMain.bat"文件，就会把代码合并到 main 目录中，后续所有代码都以这种补丁的方式提供，只把修改或新增的文件列出来。读者可以使用 Beyond Compare 工具(请读者自己安装，使用方法见"附录 B Beyond Compare 工具介绍")的文件夹比较功能，比较章节补丁的目录和 main 目录的差异，看看笔者修改的地方。本章的代码只涉及了 4 个文件，如图 2.3-8 所示。

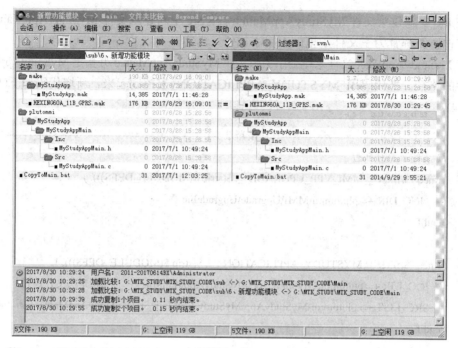

图 2.3-8

2.4 第一个程序

1. Hello Word

前面已经介绍过，MTK 中的编程语言是使用 C 语言，如果读者完全没有学过 C 语言，那得先恶补一下，如果有一点点基础，那也可以摸着石头过河，但扎实的 C 语言基础依旧是快速入门的保障。跟所有的编程入门程序一样，本书的第一个程序，也使用万能的"hello world!"。我们所要实现的功能就是在模拟器上显示一行"hello world!"。

在上一章节中，我们添加了自己的模块"mystudyapp"，但这个模块中添加的头文件和源文件都是空的，从这一章节开始，我们要不断地丰富模块的功能。

在 Source Insight 中打开 MyStudyAppMain.c 文件(习惯在 VS2008 中编程的，也可以直接在 VS2008 中写代码)，编写代码如下(代码清单 2.4-1)：

---------------------------------------代码清单 2.4-1---
```
void mtk_helloworld(void)
{
    printf("Hello World!");
}
```

这就是 C 语言代码，程序定义了一个 mtk_helloworld 函数，使用 printf 函数输出"Hello world!"。那要怎么执行这个函数呢？读者肯定能想到把它放到 main 函数里面调用，但在

MTK 系统里，没有可以供我们填充代码的 main 函数。我们暂时先不管，在 VS2008 中打开模拟器工程，以调试模式(按 F5)把模拟器运行起来，模拟器如图 2.4-1 所示。

图 2.4-1

图 2.4-1 左侧是模拟器界面，右侧是模拟器的控制台信息。模拟器界面上半部分是显示屏，下半部分是按键区域，图上标注了四个比较常用的按键名称，分别为：KEY_LSK(左软键，也叫确认键)、KEY_RSK(右软键，也叫返回键)、KEY_SEND(拨号键，主要功能是拨打电话)、KEY_END(开关机键，长按开机或关机，短按退回到 idle(待机)界面)，其次还有上下左右四个方向键、音量控制键、数字键。读者可用鼠标点不同的按键，试试不同按键的功能。我们单击左软键(KEY_LSK)会进入如图 2.4-2 所示界面。

图 2.4-2

这里进入的是主菜单界面，现在我们把这个主菜单的入口函数改一下，单击左软键不进入主菜单，而是执行 mtk_helloworld 函数。

在 Source Insight 中打开 MainMenu.c 文件(路径为：plutommi\mmi\MainMenu\MainMenuSrc，除非在 Source Insight 中会出现多个相同的文件名，否则不会再注明文件路径)。找到 goto_main_menu 函数，这个函数就是主菜单的入口函数，新增代码如下(代码清单 2.4-2)：

```
------------------------------------------------代码清单 2.4-2------------------------------------------------
/*省略部分代码*/
#if defined(__MYSTUDY_APPLICATION__)
#include "MyStudyAppMain.h"
#endif
/*省略 goto_main_menu 函数注释说明*/
void goto_main_menu(void)
{
    /*----------------------------------------------------------------*/
    /* Local Variables                                                */
    /*----------------------------------------------------------------*/


    /*----------------------------------------------------------------*/
    /* Code Body                                                      */
    /*----------------------------------------------------------------*/
#if defined(__MYSTUDY_APPLICATION__)
    mtk_helloworld();
    return;
#endif
/*省略部分代码*/
}
```

新增代码都用 __MYSTUDY_APPLICATION__ 宏包含起来了，首先使用#include 包含头文件"MyStudyAppMain.h"，头文件"MyStudyAppMain.h"里面使用 extern 关键字引用了 mtk_helloworld 函数，然后在 goto_main_menu 函数里面调用 mtk_helloworld 函数，执行完后就 return。MyStudyAppMain.h 文件中的代码如下(代码清单 2.4-3)：

```
------------------------------------------------代码清单 2.4-3------------------------------------------------
#ifndef __MYSTUDYAPPMAIN_H__
#define __MYSTUDYAPPMAIN_H__
#if defined(__MYSTUDY_APPLICATION__)


extern void mtk_helloworld(void);
#endif
#endif
```

再次运行模拟器，鼠标单击左软键，此时已经无法显示主菜单界面了，而且界面上没有任何变化，并没有输出我们想要的"Hello world!"，但是 mtk_helloworld 函数确实已经执行了(读者可通过打断点的方式查看)，只不过 printf 函数并不能把字符输出到 MTK 模拟器的屏幕上，在真机设备上也不会有任何显示，只能输出到模拟器的控制台信息里面。如图 2.4-3 所示。

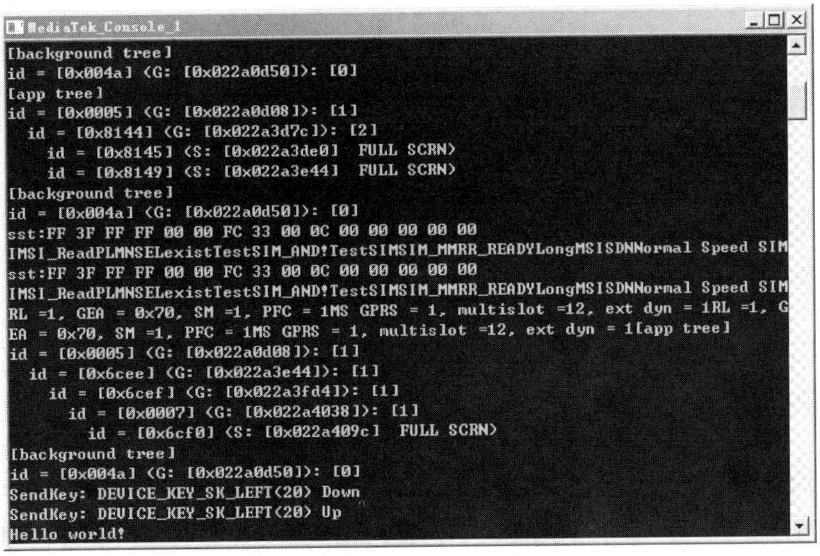

图 2.4-3

这并不是我们想要的效果，我们要把字符输出到屏幕上，需用另一个函数 gui_print_text 代替 printf，并且在 MTK 设备屏幕上显示字符，还要指定坐标、颜色、字体等，修改 mtk_helloworld 函数代码如下(代码清单 2.4-4)：

--代码清单 2.4-4--

```
#if defined(__MYSTUDY_APPLICATION__)
#include "gui_themes.h"

void mtk_helloworld(void)
{
    gui_set_text_color(UI_COLOR_BLACK);        /*设置字符颜色为黑色*/
    gui_move_text_cursor(10, 10);              /*移动字符显示坐标为(10, 10)*/
    gui_set_font(&MMI_large_font);             /*设置字符显示的字体*/
    gui_print_text("Hello world!");            /*在屏幕上打印字符*/
}
#endif
```

再次运行模拟器，单击左软键，屏幕上依旧没有任何变化，连控制台也不输出"Hello world!"了。其实我们还少做了一步操作，那就是刷新屏幕。在 MTK 系统中对屏幕显示内容做的任何更改，都必须刷新屏幕，否则依旧显示的是上一次的内容。在 mtk_helloworld 函数最后面，添加一行代码：

```
gui_BLT_double_buffer(0, 0, UI_DEVICE_WIDTH, UI_DEVICE_HEIGHT);
```
运行模拟器，此时终于可以看到屏幕上多了点东西了，但依旧不是"Hello world!"，而是显示的乱码，如图 2.4-4 所示。

图 2.4-4

这是因为 MTK 系统的屏幕上显示的任何字符，都必须是 UCS2 编码格式，而我们上面输入的字符是 ASC 编码，所以无法在屏幕上正常显示，接下来我们只要做一件事情，那就是把"Hello world!"转换成 UCS2 编码。修改代码如下(代码清单 2.4-5)：

--代码清单 2.4-5--

```
#if defined(__MYSTUDY_APPLICATION__)
#include "gui_themes.h"

void mtk_helloworld(void)
{
    U8 str_asc[] = "Hello world!", str_ucs2[32] = {0x00};

    gui_set_text_color(UI_COLOR_BLACK);/*设置字符颜色为黑色*/
    gui_move_text_cursor(10, 10);      /*移动字符显示坐标为(10, 10)*/
    gui_set_font(&MMI_large_font);    /*设置字符显示的字体*/

    mmi_asc_to_ucs2((CHAR*)str_asc, (CHAR*)str_ucs2);
    gui_print_text(str_ucs2);          /*在屏幕上打印字符*/

    gui_BLT_double_buffer(0, 0, UI_DEVICE_WIDTH, UI_DEVICE_HEIGHT);
}
#endif
```

再次运行模拟器，终于能看到我们想要的效果了，如图 2.4-5 所示。

图 2.4-5

以上代码，就是 MTK 中的一个 hello world 程序。总结一下，在 MTK 设备的屏幕上显示字符，要做如下几步操作：

(1) 设置字符显示的颜色、坐标、字体，这三个动作不分先后顺序，一次设置，永久有效，直到下一次设置的时候才会改变原来的设置。

(2) 打印 UCS2 编码格式的字符。

(3) 刷新屏幕。

其实以上代码还有一种比较简洁的写法，直接把 gui_print_text("Hello world!")改成 gui_print_text(L"Hello world!")，在字符串常量的前面添加一个大写的"L"，可以把字符串转换成 UCS2 编码，读者可以自己尝试一下，最终的运行效果与前面是一样的，代码如下(代码清单 2.4-6)：

---------------------------------------代码清单 2.4-6---

```
#if defined(__MYSTUDY_APPLICATION__)
#include "gui_themes.h"

void mtk_helloworld(void)
{
    gui_set_text_color(UI_COLOR_BLACK);/*设置字符颜色为黑色*/
    gui_move_text_cursor(10, 10);/*移动字符显示坐标为(10, 10)*/
    gui_set_font(&MMI_large_font);/*设置字符显示的字体*/

    gui_print_text(L"Hello world!");/*在屏幕上打印字符*/

    gui_BLT_double_buffer(0, 0, UI_DEVICE_WIDTH, UI_DEVICE_HEIGHT);
}
#endif
```

现在我们的程序在模拟器上已经编译通过并且能够在模拟器上正常运行了。接下来我们再用 make 命令编译试试，如果前面已经对整个工程执行了 make new，并且编译通过，那么这里只要 remake 我们所修改的源文件(.c)所对应的模块就行了，我们的代码只涉及两个源文件，在 MyStudyAppMain.c 文件中编写 mtk_helloworld 函数，然后放到 MainMenu.c 文件中调用执行。MyStudyAppMain.c 文件是我们自己添加的，添加在 MyStudyApp.mak 文件中，对应 mystudyapp 模块。那么 MainMenu.c 文件对应哪个模块呢？我们在 Source Insight 工具中按下快捷键"Ctrl+/"，在弹出的对话框中输入 MainMenu.c，然后单击"Search"按钮，如图 2.4-6 所示。

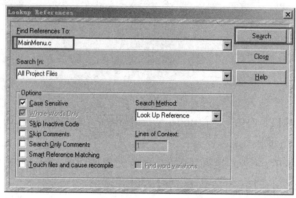

图 2.4-6

搜索完成后，可以看到如图 2.4-7 所示的搜索结果，在结果中可以看到有个 Mmi_app.mak 的 mak 文件，那么 MainMenu.c 文件就对应 mmi_app 模块。如果有的人搜索无法得出这个结果，则需要 rebuild 一下 Source Insight 工程。

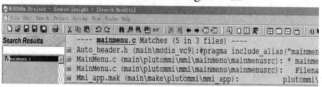

图 2.4-7

修改的两个源文件对应的模块都找到了，那我们需要执行的编译指令为 make r mystudyapp mmi_app，但遗憾的是，编译 mmi_app 模块报错了，如图 2.4-8 所示。

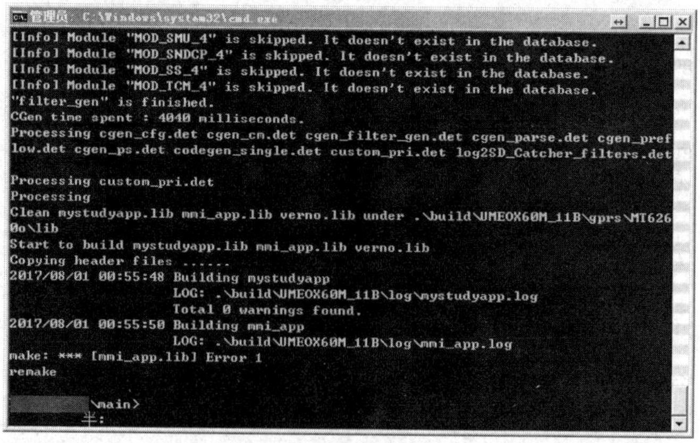

图 2.4-8

我们到 build\HEXING60A_11B\log 里面找到 mmi_app.log 文件并打开(推荐使用 UltraEdit 工具)，定位到文件最后，内容如图 2.4-9 所示。我们搜索"Error:"，可以查找到如红框中标记的错误提示信息，大致意思为：在 mainmenu.c 文件中 2828 行包含的头文件 MyStudyAppMain.h 找不到，出错的代码为#include"MyStudyAppMain.h"。MyStudyAppMain.h 文件在目录 plutommi\MyStudyApp\MyStudyAppMain\Inc 中，这个目录我们只在 MyStudyApp 模块中添加了，并没有在 mmi_app 模块中添加，所以在 mmi_app 模块中使用#include，是无法包含文件的。

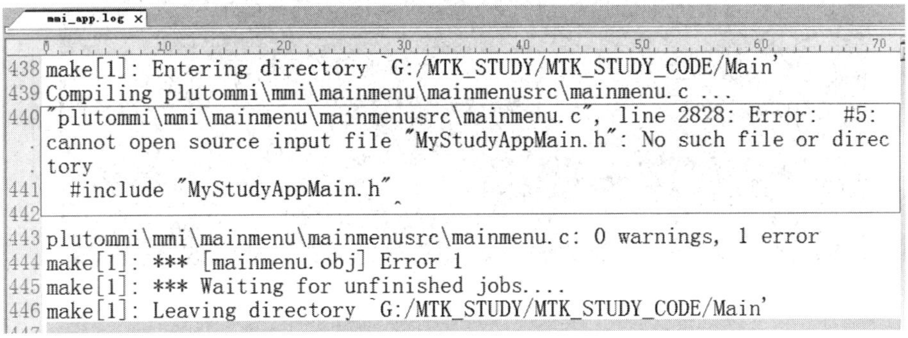

图 2.4-9

这个问题有三种解决方法。我们包含这个头文件的目的就是为了调用 mtk_helloworld 函数，所以第一种方法是去掉#include "MyStudyAppMain.h"语句，改为直接使用 extern 引用 mtk_helloworld 函数，这种方式仅适用于只需要引用外部函数，如果头文件里还包含了其他的定义，比如宏定义、结构体类型定义，此时就不能使用 extern 了，必须使用#include 引用头文件。第二种方法就是在#include 里面写上头文件的路径，这种方式也能解决头文件路径找不到的问题，但书写比较麻烦。最后一种是在 mmi_app 模块里面添加 MyStudyAppMain.h 头文件的路径，打开 mmi_app.mak 文件，在最后面添加代码如图 2.4-10 所示。本书推荐使用这种方式，但切记千万不能再把源文件(.c)重复添加，否则也会编译错误。

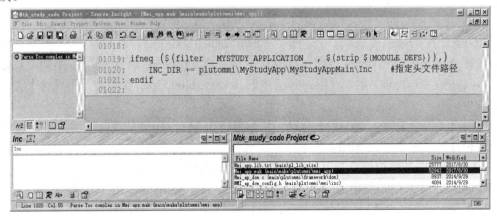

图 2.4-10

重新执行 make r mystudyapp mmi_app 指令，此时不再报错，编译完成后，显示结果如图 2.4-11 所示，并且在 build\HEXING60A_11B 目录下生成 HEXING60A_11B_PCB01_gprs_MT6260_S00.HEXING60A_11B_SW.bin 文件夹。

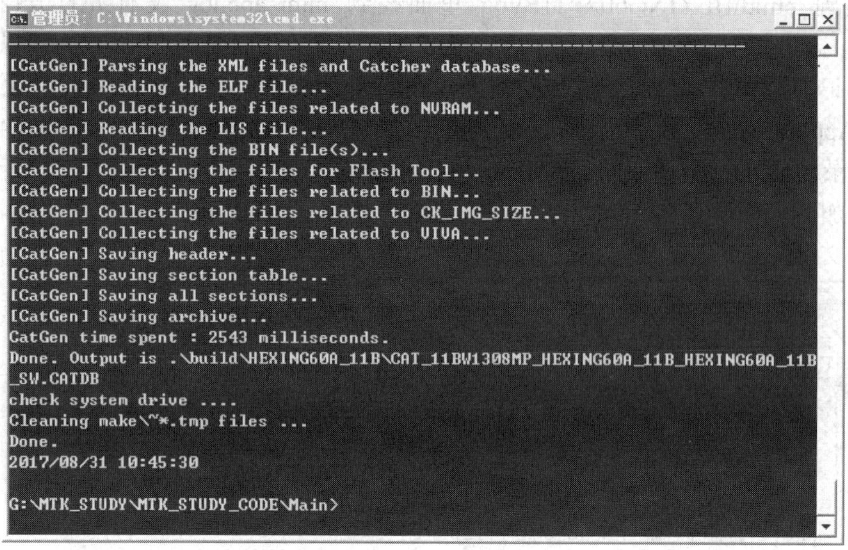

图 2.4-11

2. 变量

在上面的 hello world 程序中，定义了两个 U8 类型的数组，数组里面存储的内容为字符串。在 C 语言里面也有存储字符串的数据类型，就是 char 型数组，char 类型是带符号的，还有一种无符号的 unsigned char。前面已经说过，MTK 的代码都是 C 语言编写的，从这里可以判断 U8 类型实际上就是 char 或者 unsigned char 类型，那为什么命名为 U8 呢？

注意 "unsigned"（无符号）的第一个字母就是 "U"，U8 就是 unsigned char 类型。那 "8" 又表示什么含义呢？学过 C 语言的都知道，一个字符类型不管是有符号还是无符号都是占 1 个字节(byte)，1 个字节有 8 位(bit)，这里的 "8" 就表示位数，与 8 整除就是字节数。所以 "U8" 可以理解为无符号的 8 位数据类型，或无符号的 1 字节数据类型，在 C 语言里面就对应 unsigned char 类型。

MTK 里面所有的无符号数据类型都是以 "U"+位(bit)数来表示的。比如 U16 就是无符号 16 位(2 字节)数据类型，对应 C 语言中的 unsigned short(32 位编译器，下同)类型；U32 就是无符号 32 位(4 字节)数据类型，对应 C 语言中的 unsigned int 类型；还有 U64 就是无符号 64 位(8 字节)数据类型，对应 C 语言中的 unsigned long long 类型，只不过这个类型用得比较少。

既然有无符号数据类型，那么有符号数据类型在 MTK 中又是如何表示的呢？在 C 语言中有符号的字符类型为 char，实际上前面还省略了一个关键字 signed(有符号)，所以，MTK 中有符号的数据类型都是以 "S"+位(bit)数来表示的。比如 S8 就是有符号 8 位(1 字节)数据类型，对应 C 语言中的 char 类型；S16 就是有符号 16 位(2 字节)数据类型，对应 C 语言中的 short 类型；S32 就是有符号 32 位(4 字节)数据类型，对应 C 语言中的 int 类型。

这些数据类型都可以在 MMIDataType.h 中查看到，实际上就是在 C 语言数据类型的基础上用 typedef 定义了一个别名。这样做的好处是，我们不用去关心编译器是 16 位的还是 32 位的，从变量的数据类型，就能一目了然地看出它占几个字节，取值范围是多少。比如

U16 占两个字节，取值范围为 0~0xFFFF(65535)；S8 占 1 个字节，取值范围为-128~127。

值得注意的是，C 语言中的同一个数据类型，却在 MTK 中有多个别名，比如 char 的别名有 S8、kal_char、kal_int8，其他的类型也有这种情况，而且他们的 typedef 定义出现在多个文件里面，可以在 kal_general_types.h 文件中查看以"kal_"开头的数据类型定义。不同的头文件引用范围不同，MMIDataType.h 头文件仅用 MMI 层模块的代码，在驱动层是无法引用的，但驱动层可以引用 kal_general_types.h 头文件，所以驱动层的数据类型都是"kal_"开头的。

不管 MTK 的数据类型如何命名，它始终是万变不离其宗，我们在 MTK 中写代码的时候，既可以用 MTK 的重定义数据类型名称，也可以用 C 语言原生态的数据类型名称。使用 MTK 的数据类型可以与原生代码统一风格，方便理解，但是如果跨平台移植，比如移植到非 MTK 平台上，就会出现数据类型不兼容，很多类型都找不到定义。而使用 C 语言的数据类型，就不会存在这个问题，但书写会比较繁琐，而且 MTK 中很多函数接口的参数都是用的 MTK 的数据类型，有时需要进行强制类型转换。

总之，不管使用 MTK 的数据类型还是 C 语言原生态的数据类型都没有问题，它们各有各的优点和缺点，如果确定自己的代码只用于 MTK 平台上，那就遵循 MTK 的规则，如果需要跨平台移植，就用标准的 C 语言编码方式。本书中的代码会比较偏向于使用 MTK 的数据类型。

2.5 下 载 和 调 试

如果读者自备了 MTK 硬件开发套件，则可以自己动手演示这一章节的内容，如果没有自备 MTK 硬件开发套件，则只需了解真机上的下载和调试步骤就可以了，以后有了开发套件，再来查看这一章节的内容。但是对于模拟器上的 catcher 工具使用方法，还是建议读者提前掌握。本章节采用 6260M 的儿童定位智能手表来作演示，该项目的名称为 UMEOX60M_11B。

2.5.1　FlashTool 下载

我们在 MTK 上写的程序，最终都要烧录到 MTK 硬件设备中去运行，这个烧录的过程就称之为下载。在 2.4 节中我们写了一个 helloworld 程序，虽然只是在模拟器上运行，但我们使用 make 命令编译通过，已经生成了二进制下载文件，这一章节中我们把它下载到真机设备中。

make 命令编译生成的文件大部分都在 build 目录下，这个目录中有一个与项目名称一模一样的文件夹 UMEOX60M_11B，如果编译了多个项目，就会存在多个项目文件夹中，其中包含了各自项目的生成文件，进入 UMEOX60M_11B 目录中可以看到如图 2.5-1 所示的内容，其中有一个文件夹名为 UMEOX60M_11B_PCB01_gprs_MT6260_S00.MAUI_11B_W13_08_MP_V9.bin，这个文件夹里面就是编译生成的二进制下载文件，如果代码编译出错，这个文件夹是不会生成的。

图 2.5-1

1. 驱动安装

MTK 系统下载，需要使用电脑 USB 串口工具连接 MTK 硬件设备，所以在下载之前，我们需要先安装 USB 串口驱动。解压"USB 串口驱动.zip"压缩包，这个目录下面总共有两个可执行程序需要安装。

执行 MTK_USB_Driver_V1.1121.0\ComPortDriver\InstallDriver.exe 文件，安装完成后出现如图 2.5-2 所示界面，这里需要特别注意提示框中显示的字符"Install USB Driver for Microsoft Windows 7 Ultimate Edition(64-bit) successfully"，其中"Microsoft Windows 7 Ultimate Edition(64-bit)"就是我们电脑安装的系统名称"WIN7 64 位旗舰版"，这个名称必须与系统一致。如果提示为"WIN7 64 位旗舰版"，但我们实际上安装的是"WIN XP"系统，那就说明驱动安装失败了。笔者曾经遇到过这样的问题，但没有找到很好的解决方法，只能换一个操作系统重装。

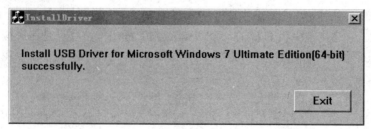

图 2.5-2

执行 MTK_USB_Driver_V1.1121.0\ModemPortDriver\ModemInstaller.exe 文件，安装完成后出现如图 2.5-3 所示界面，此处同样要注意提示中系统名称是否与自己电脑安装的系统一致。

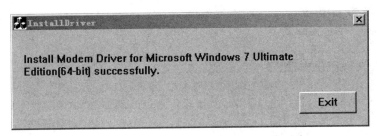

图 2.5-3

执行 PL-2303 Driver Installer\PL-2303 Driver Installer.exe 文件,这个安装没什么注意的地方,全部按照默认安装即可。

2. Flash Tool 下载

驱动安装成功后,就可以用 MTK 下载工具——FlashTool 下载。FlashTool 是单串口下载工具,每次只能下载一个设备,另外还有一个多串口下载工具,可以同时下载多个设备,主要用于工厂生产,在这里不作介绍。FlashTool 不需要安装,本书使用的版本是"FlashTool_V5.1640.00.zip",解压之后双击"Flash_tool.exe"文件就可以运行。软件运行的界面如图 2.5-4 所示。

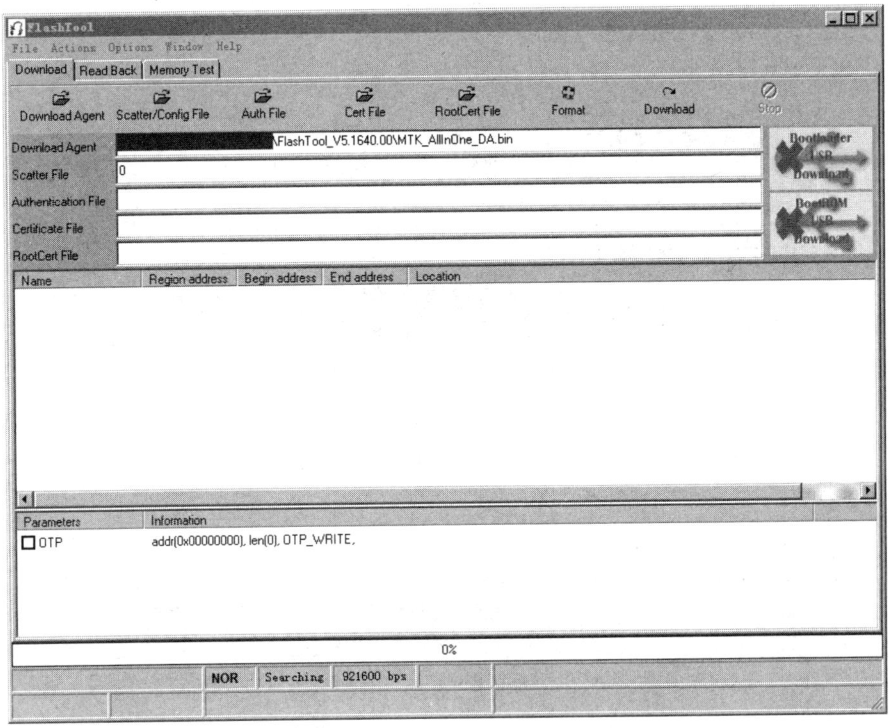

图 2.5-4

单击"Scatter/Config File"按钮,在弹出的文件选择框中,选择 MTK 源码中的 \build\UMEOX60M_11B\UMEOX60M_11B_PCB01_gprs_MT6260_S00.MAUI_11B_W13_08 _MP_V9.bin\UMEOX60M_11B_BB.cfg 文件,如图 2.5-5 所示,然后单击"打开"按钮,加载下载文件,如图 2.5-6 所示。

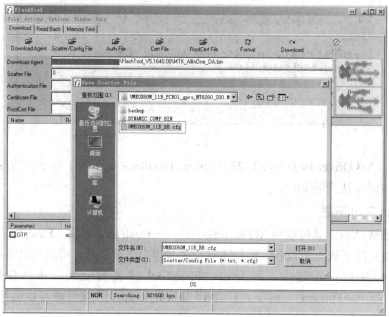

图 2.5-5

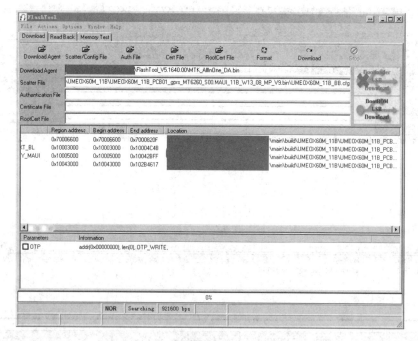

图 2.5-6

单击"Options"菜单，会弹出一系列的子菜单，如图 2.5-7 所示，这里面有几个选项是我们经常会接触到的，在这里给大家讲解如下。

(1) COM Port：使用串口线下载时需要选择这个选项的串口列表中对应的串口，本书不作重点介绍。

(2) USB Download/Readback：使用 USB 串口也就是 USB 数据线下载，本书下载程序都是用 USB 数据线，所以必须选择这个选项。

(3) USB Download Without Battery：下载系统时不需要连接电池，如果不选此项，则设备没有电池不能下载。

(4) Format FAT：格式化 MTK 设备，同程序面板中的 Format 按钮功能一样，这个功能要慎用，后面再作详细介绍。

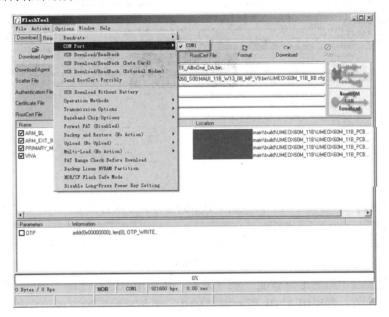

图 2.5-7

此处我们只需要选择"USB Download/Readback"选项，选中之后会在菜单前面有个勾，如图 2.5-8 所示。

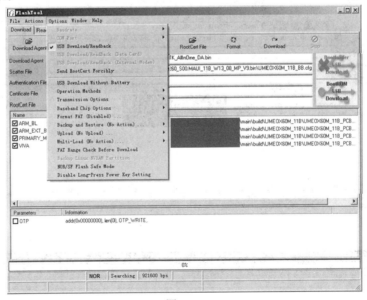

图 2.5-8

单击 Download 按钮，此时其他的按钮都会变成灰色，只有一个 Stop 按钮可以操作。如图 2.5-9 所示。

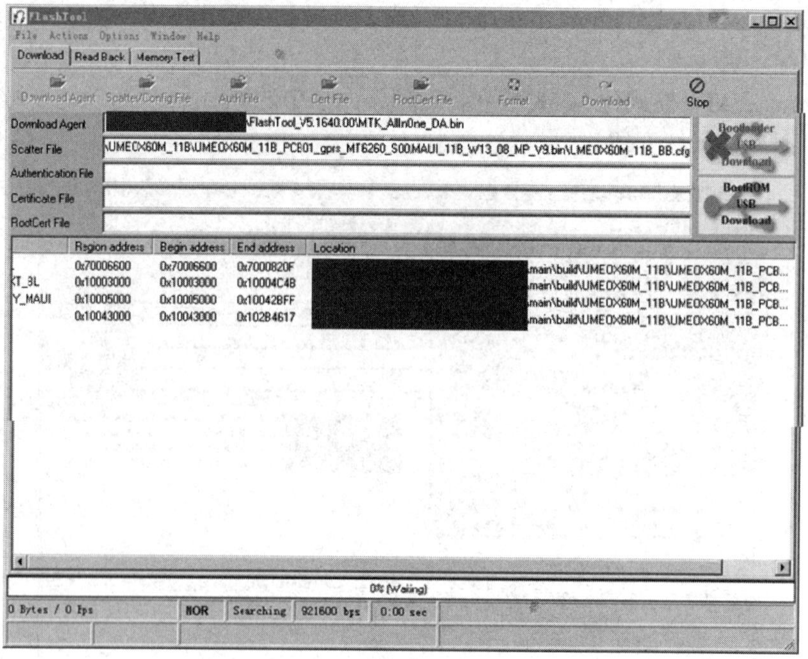

图 2.5-9

把本书配套的"儿童定位智能手表"关机，然后用 USB 线连上电脑，如果出现如图 2.5-10 所示的界面，则说明 USB 驱动安装成功，系统已经开始下载，底部出现的 COM4 是对应的 USB 串口号。如果没反应，则说明 USB 驱动安装失败。

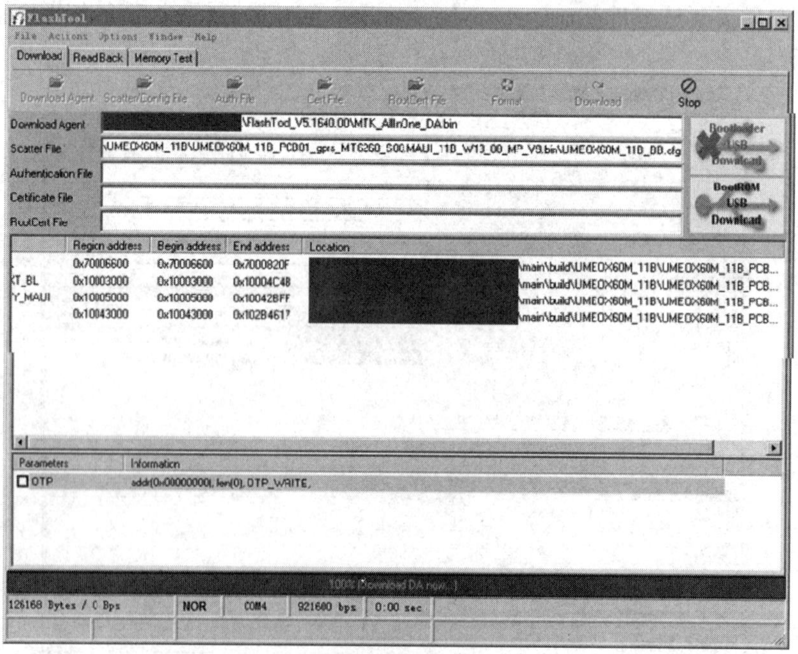

图 2.5-10

在下载的过程中，会出现好几次颜色不同的进度条，比如图 2.5-11 所示。但如果弹出一个绿色的圆圈，如图 2.5-12 所示，则说明下载完毕。

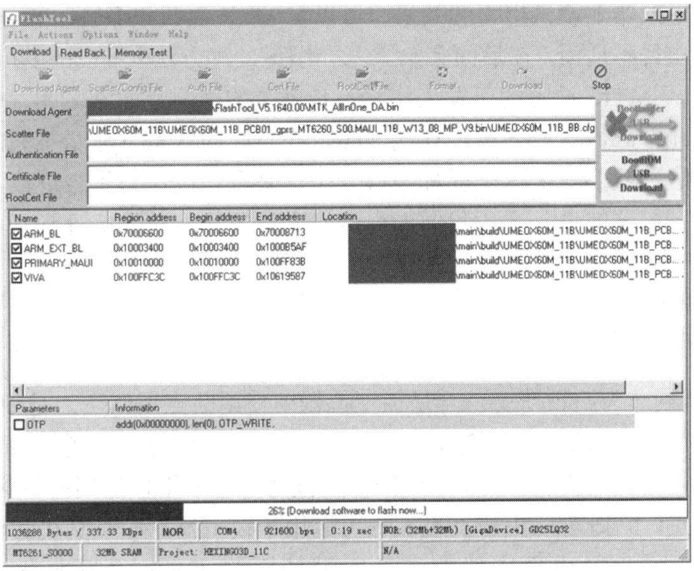

图 2.5-11

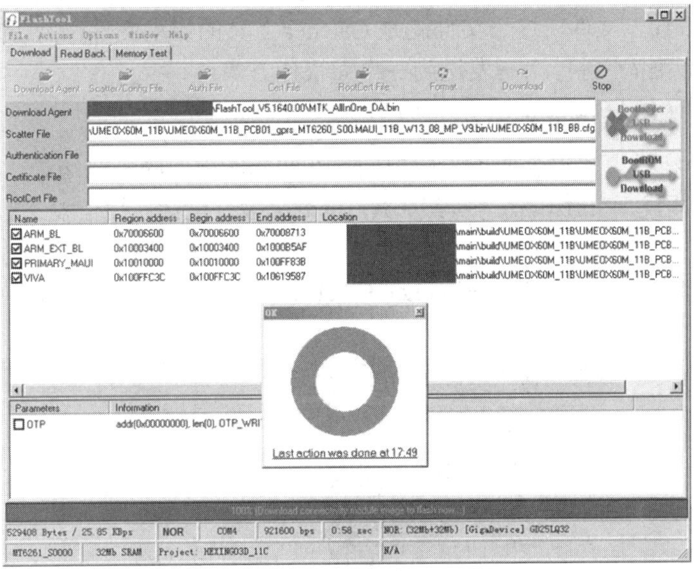

图 2.5-12

下载完成后，拔掉 USB 线，断开设备与电脑的连接，长按手表左上角的按键就可以开机了。但是很遗憾，开机后并没有出现类似于模拟器所示的界面，屏幕上只显示一个"守护"字符，其他没有变化，如图 2.5-13 所示。这是因为硬件设备的屏幕与驱动代码不匹配。我们在 UMEOX60M_11B_GPRS.mak 文件中配置屏幕尺寸为 240×320(MAIN_LCD_SIZE = 240×320)，而实际的硬件屏幕尺寸为 64×48，所以无法正常显示，也就是说本书关于屏幕显示的大部分内容都将只能在模拟器上看效果，但这并不意味着模拟器上的效果无法在真机上显示，只要

图 2.5-13

屏幕驱动支持，模拟器上调试出来的效果，可以在 MTK 真机设备上一模一样地显示出来。不过对于本书的项目不适用，在后面会给大家介绍另一种方法，可以改变屏幕的显示内容。如果读者今后从事 MTK 设备的研发工作，不带屏幕的产品是非常多的，但是不带屏幕的设备一般都会配几个灯，只要开机，灯就会亮，在不同的状态下显示不同的灯效，比如摩拜单车上的锁。这里我们就把这个屏幕当做一个灯来看，仅仅为了告诉读者这个设备已经开机了。但在没有屏幕的设备上，我们如何能确定设备什么时候执行到了自己所写的代码呢？这就是下一节要讲解的内容。

2.5.2　Catcher 调试

MTK 代码调试分为两种情况，一是在模拟器中使用断点调试，二是使用 Catcher 工具调试。模拟器上的断点调试比较简单，通过打断点，跟踪代码执行逻辑，查看程序运行时内存中的变量值，这种方式属于 VS2008 的应用技巧，本书不作专门的讲解。Catcher 工具调试稍微麻烦一点，它既可以用于真机，也可以用于模拟器，而且真机和模拟器上的应用并不相同。Catcher 工具也不需要安装，本书使用的 Catcher 版本是 V3.1532，解压"Catcher_V3.1532.00.zip"压缩包，在"Catcher_V3.1532.00\[3.1532.00]Catcher"目录下可以看到"Catcher.exe"文件，这个文件就是 Catcher 工具的启动文件，为了方便使用，可以创建快捷方式到桌面(右键单击→发送到→桌面快捷方式)。

1．模拟器中使用 Catcher 工具

Catcher 工具在模拟器中主要用于模拟硬件在实际应用中的一些场景，比如充电、电池电量变化、耳机插拔、SIM 卡信号等级、AT 指令、拨打电话、短信发送等。使用 Catcher 工具步骤如下：

(1) 运行模拟器，点击"Tool→Launch Catcher and NS"菜单，弹出如图 2.5-14 所示提示框。

图 2.5-14

(2) 单击"确定"按钮，在弹出的对话框中，单击右侧的省略号"……"按钮，进入 Catcher 工具所在的目录，选择"Catcher.exe"文件，单击"OK"按钮(如图 2.5-15 所示)。

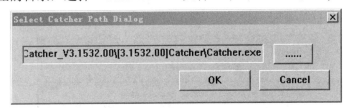

图 2.5-15

当我们编译一个新的模拟器第一次使用 Catcher 工具时，都要进行上述 Catcher 工具的路径设置，设置好后，Catcher 工具的路径会保存在 MoDIS_VC9\MoDIS\Debug\MoDIS.ini 文件中，该文件中保存的 Catcher 工具默认路径为：CATCHER = T:\[3.0852.00]Catcher\Catcher.exe。如果这个文件不丢失，下次重新运行模拟器不需要再次设置路径。如果不小心把路径设置错了，则无法通过上述方法重新设置 Catcher 路径，我们可以直接修改这个文件，设置 CATCHER 属性的值，即 Catcher.exe 文件的完整目录。比如 Catcher 目录为"E:\Catcher_V3.1532.00\[3.1532.00]Catcher"，那么直接设置"CATCHER = E:\Catcher_V3.1532.00\[3.1532.00]Catcher\Catcher.exe"就可以了。接下来，我们再次单击模拟器上的菜单"Too→Launch Catcher and NS"菜单，就可以启动 Catcher 工具了(如图 2.5-16 所示)。

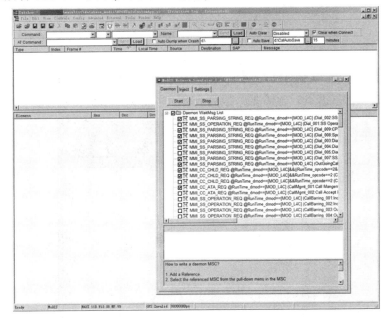

图 2.5-16

在模拟器中打开 Catcher，还会启动另外一个程序：MoDIS_VC9\NetSimScript.msc。如图 2.5-16 所示的 MoDIS Network_Simulatre_1 窗口，这个程序包含在 MTK 源码包中，但无法单独运行。在 Catcher 窗口顶部有一排工具栏，但在模拟器中大部分都无法使用，只有一个标签为"AT Command"的输入框可以在模拟器中模拟 AT 指令，在后面讲 AT 指令的时候会给大家讲解。我们在模拟器中使用的 Catcher 功能大部分都在 MoDIS Network_Simulatre_1 窗口中操作，它能够模拟一些真机的使用场景。

(3) MoDIS Network_Simulatre_1 有三个标签栏，如图 2.5-16 所示，分别为：Daemon、Inject、Settings。我们在"Daemon"标签栏中点击"Start"按钮，让模拟器与 Catcher 工具建立连接。如图 2.5-17 所示。

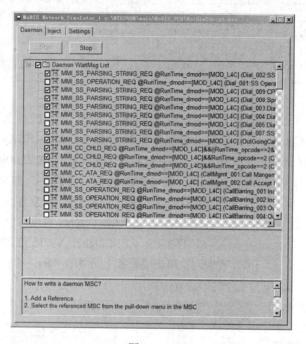

图 2.5-17

然后切换到"Inject"标签栏，这个标签栏中就有一个"Test Case List"(测试用例列表)，这个列表默认是全部展开的，笔者把二级列表合上了，如图 2.5-18 所示。

图 2.5-18

下面给大家介绍几个比较常用的测试用例，其他的测试用例读者可以自己去尝试。我们展开 IncomingCall 列表，这个用例是用来测试模拟器来电的情况。鼠标右键单击第一项 IncomingCall_001，然后鼠标左键单击 Inject(如图 2.5-19 所示)，观察模拟器的变化(如果模拟器灭屏了，点击任意按键可唤醒)，是否发现模拟器有电话打进来了，来电号码为 10000，这个号码是可以修改的。当我们选中 IncomingCall_001 时，在图 2.5-19 所示的窗口中有个 Parameter 参数列表，可以修改 Number 的值为任意号码，读者可以自己尝试。在模拟器的来电界面，我们可以单击不同的按键，选择接听或者挂断。在软件上代码执行的流程与真机上几乎是完全一样的。这里需要提醒一下，如果没有单击"Daemon"标签栏中的"Start"按钮，在模拟器上是无法接听或挂断来电的。

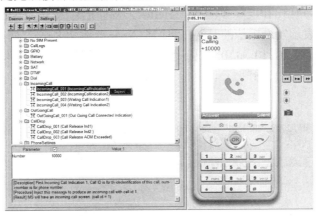

图 2.5-19

在"Test Case list"(测试用例列表)中所有测试用例的使用方法都是右键单击再选择 Inject，在后面的讲解中会直接简写成"Inject IncomingCall→IncomingCall_001"，或者"执行 IncomingCall->IncomingCall_001"，读者应能明白是什么意思，并知道如何操作。

既然可以模拟来电(电话打进来)，当然也可以模拟去电(电话打出去)。我们在模拟器上直接点击数字按键，输入 10086(如图 2.5-20 所示)，然后单击拨号键(图中红色箭头指向的按键)，在出现的 SIM 卡选择界面中再次单击拨号键，此时就直接进入了通话界面。

图 2.5-20

在单击了"Daemon"标签栏中的"Start"按钮后,模拟器上拨号是可以直接接通的。如果想要测试去电的界面,读者可单击 Daemon 标签栏中的"Stop"按钮,先断开与模拟器的连接,然后再拨号,此时就能看到模拟器正在拨号,但没有接通的情况。如果想要接通电话,就再次单击"Daemon"标签栏中的"Start"按钮,然后切换到"Inject"标签栏,执行 OutGoingCall→OutGoingCall_001(Out Going Call Connected Indication)操作,如图 2.5-21 所示。如果想要挂断电话,可以在模拟器中单击"end"键,也可以执行 CallDrop→CallDrop_001 (Call Release Ind1)。

除了打电话的功能外,我们的手机还拥有发送短信的功能,在 Catcher 中也能进行模拟。找到 SMS 列表中的 SMS_008(MT SMS),用鼠标单击选中它,在下面的 Parameter(参数)列表中,Content 表示短信的内容,默认是"Hello",我们把它改为"good good study, day day up!";后面的参数表示发送短信的时间,我们把 Year 改成 2080,其他都为 1,再把发送者号码 senderNumber 改为 110。110 在未来给我们的模拟器发了一条短信,要我们"好好学习,天天向上!"。参数设置好后,再执行 Inject 操作,或者单击工具栏中的"SMS"按钮,如图 2.5-22 所示。

图 2.5-21

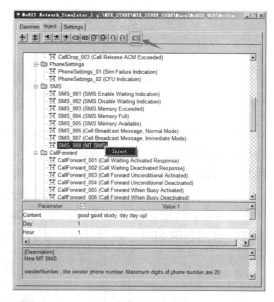

图 2.5-22

执行 SMS→SMS_008(MT_SMS) 之后,模拟器上就可以看到有新短信提醒(如图 2.5-23 左图)。我们点击左软键(view)进入中间图示界面,可以看到发送短信的号码为 110,发送时间为"01.01.2080 05:01",但是我们在前面设置的时间小时为 1,为什么这里会显示 5 呢?这跟系统中默认的时区有关系,比如你在中国给一个美国朋友发送一条短信,因为时差的关系,如果对方收到短信显示的是北京时间,那肯定是不合理的。所以当系统收到短信时,会把时间转换为 MTK 设备系统设置的时区时间。继续点击左软键,进入 option→view 菜单,就可以查看短信,短信内容就是我们之前设置的。

这是在模拟器中模拟接收短信的情况,至于发送短信,则可以直接在模拟器中操作。不需要借助 Catcher 工具。

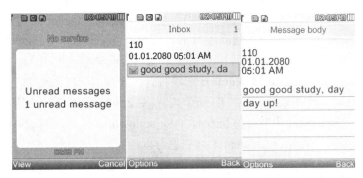

图 2.5-23

限于篇幅,这里只给大家介绍通话和短信的测试用例,还有其他的测试用例,读者可以自己尝试,比如插入耳机、拔出耳机、插入充电器、拔出充电器、满电状态、低电状态、插入 SIM 卡、拔出 SIM、SIM 无信号、SIM 满信号等。

以上内容介绍了 MoDIS Network_Simulatre_1 窗口中模拟真机使用场景的功能,这些功能可以帮助我们测试程序功能的正确性及完整性。如果模拟过程中发现程序功能错误,我们就需要调试代码,前面介绍过了,在模拟器中调试代码,主要是通过 VS2008 断点调试,另外还有一种方式就是通过 Catcher 工具打印 log 信息,通过 log 信息,我们可以查看程序的执行状态以及执行流程。

回到我们的代码中,打开文件 MyStudyAppMain.c,在 mtk_helloworld 函数中的任意位置添加一行代码:kal_prompt_trace(MOD_XDM, "-mtk_helloworld-%d--%s--", __LINE__, __FILE__)如下(代码清单 2.5-1):

---------------------------------------代码清单 2.5-1---------------------------------------

```
#if defined(__MYSTUDY_APPLICATION__)
#include "gui_themes.h"

void mtk_helloworld(void)
{
    gui_set_text_color(UI_COLOR_BLACK);          /*设置字符颜色为黑色*/
    gui_move_text_cursor(10, 10);                /*移动字符显示坐标为(10, 10)*/
    gui_set_font(&MMI_large_font);               /*设置字符显示的字体*/

    gui_print_text(L"Hello world!");             /*在屏幕上打印字符*/

    kal_prompt_trace(MOD_XDM, "-mtk_helloworld-%d--%s--", __LINE__, __FILE__);
    gui_BLT_double_buffer(0, 0, UI_DEVICE_WIDTH, UI_DEVICE_HEIGHT);

}

#endif
```

我们暂时不管这行代码什么意思，先运行模拟器，如果 Catcher 关闭了，则重新打开，并单击"Start"按钮，当与模拟器建立连接后，可以看到 Catcher 工具的"Primitive Log：Integrated"文件窗口中多了一些字符信息，这个窗口是默认就会打开的，如果不小心关闭了，可以通过 View 菜单下的 PS Integrated 选项打开。这些信息是系统的 Log 信息，对我们没用，我们单击 Catcher 工具栏中的"Clear"按钮(图 2.5-24 中红色箭头所指按钮)，把它们全部清空，然后单击模拟器中的左软键执行 mtk_helloworld 函数，可以看到模拟器上显示"Hello world!"字符，另外在 Catcher 工具中也可以找到一行 Log 信息，如图 2.5-24 所示。

图 2.5-24

这个 Log 信息就是我们新增那条语句打印出来的，该语句及 Log 信息的详细解读，会在下一节中讲解。

2. 真机中使用 Catcher 工具

在真机中使用 Catcher 工具，打开方式跟模拟器中有点不一样，需要手动双击"Catcher.exe"运行，运行后的界面如图 2.5-25 所示。图中标记了几个常用按钮，把鼠标停在对应按钮上就能看到它的名字。在后面的讲解中，会经常提到这些按钮的名称，希望大家能记住并找到它。

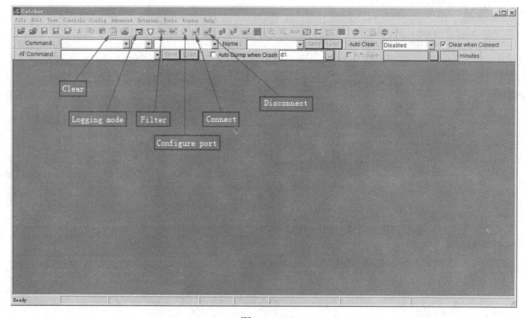

图 2.5-25

在刚打开 Catcher 工具时，图中标记的几个按钮中只有"Logging Mode"不是灰色，所以只能点这个按钮，单击它出现如图 2.5-26 所示提示框，单击"Database Path"下面名称为"Master"的输入框右侧红框标记的"..."按钮，在弹出的文件选择对话框中选中 MTK 源码目录\tst\database_classb 中的数据库文件 BPLGUInfoCustomAppSrcP_MT6260_S00_UMEOX60M_11B_SW 文件。这个目录下还有一个类似的文件，我们要选择的是以"BPLGUInfoCustomAppSrcP_"开头的文件，注意不要选错了。选好之后，单击"OK"按钮，出现如图 2.5-27 所示界面。

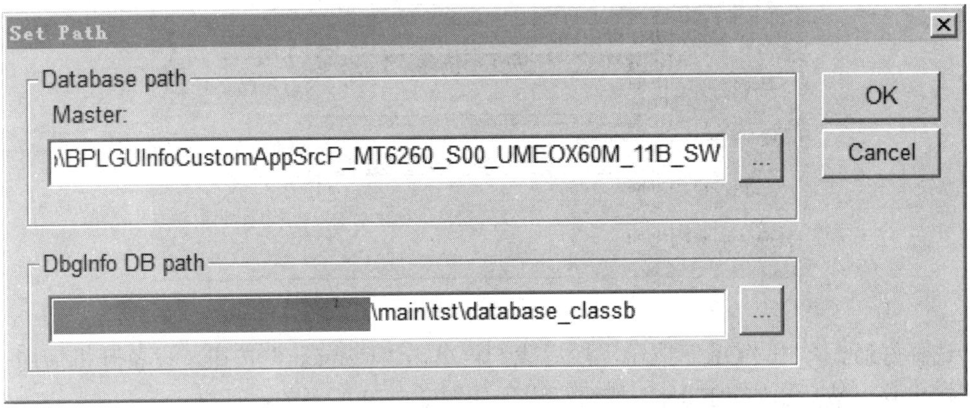

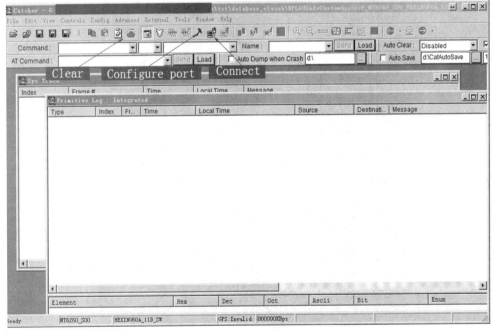

图 2.5-26

图 2.5-27

在图 2.5-27 所示的界面，可以在标题栏中看到自己选择的 database 文件名及完整路径，并且"Clear"、"Configure port"、"Connect"三个按钮不再是灰色，此时我们点击"Configure Port"按钮，出现如图 2.5-28 所示的界面，在标签名为"Port"的下拉选择框中选择"USB

Auto Connect",这个选项的功能就是使用 USB 串口连接,前面我们安装的 USB 驱动,在这里就会被用上。下面还有两个选项"None"和"COM1",都不能选。如果使用串口数据线连接 Catcher,则插入串口数据线这里会显示对应的串口,我们可以选择它即使用串口数据线连接。串口数据线和 USB 数据线的区别在于:使用串口数据线连接电脑,不会对设备充电,而且设备不开机也可以连上 Catcher;但 USB 数据线比较常见,大多数 MTK 设备都会搭配一根 USB 数据线,所以本书全部使用 USB 数据线。

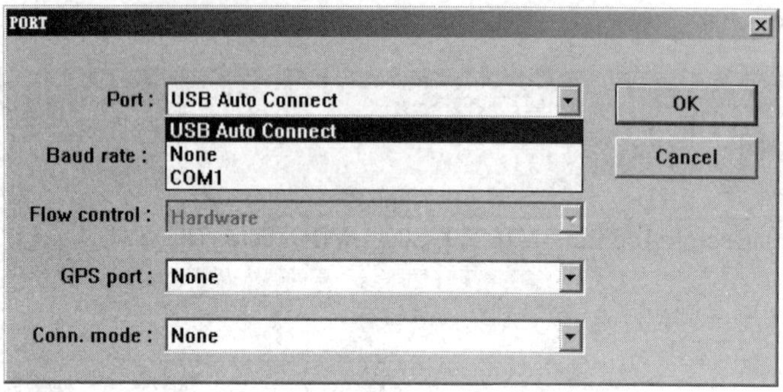

图 2.5-28

选择好后,单击"OK"按钮,然后我们再单击"Connect"按钮,设备开机后用 USB 先连接电脑,然后单击"Filter"按钮,出现如图 2.5-29 所示的对话框。

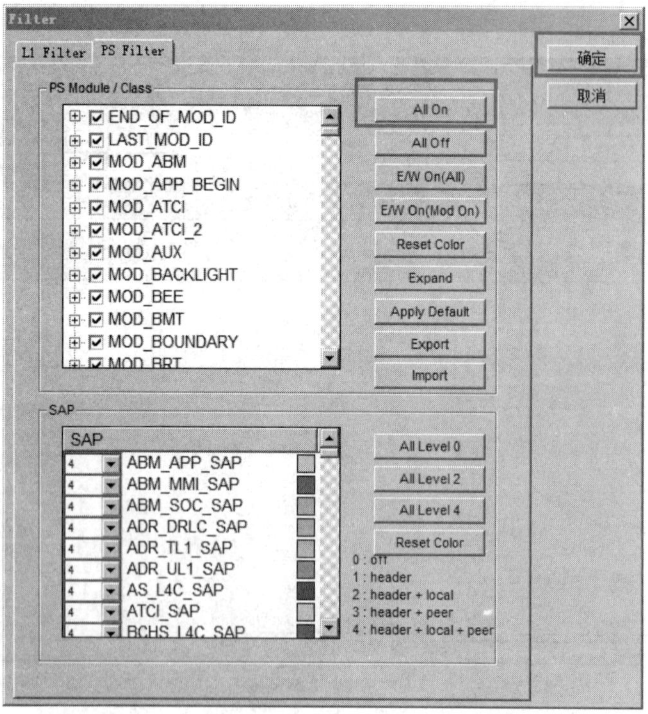

图 2.5-29

单击"All On"按钮,左侧 PS Module/Class 对话框中的所有复选框都会打上勾,然后单击"确定"按钮。如果连接成功,Catcher 工具的 Primitive Log:Integrated 窗口中会出现很多 log 信息,如图 2.5-30 所示,这些 Log 信息都是系统打印的。

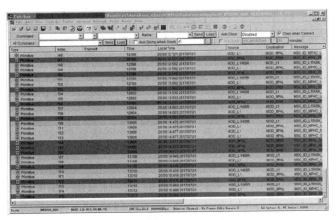

图 2.5-30

如果没有出现图 2.5-30 所示的信息，而是弹出图 2.5-31 所示的提示框，则说明设备连接失败，出现这种情况的原因有很多，比如 USB 线坏了(换根 USB 线试试)；或者电脑的 USB 接口坏了(换个 USB 插口试试)；或者手表上的 USB 接口坏了(只能修手表了)；或者驱动安装失败(卸载重装)。但如果前面使用 USB 线下载程序成功了，那上述几个问题都可以排除，此时可能是电脑上的 COM 端口被占用了。如果确保 Flash_tool 没有点击"Download"按钮，那就把电脑重启一下。

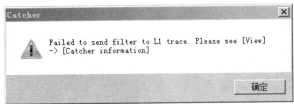

图 2.5-31

Catcher 连接成功后，我们短按一下左软键(右上角的按键)，此时，我们的 mtk_helloworld 函数应该已经执行了，而且这个函数中打印的 Log 信息应该也显示出来了，但是这么多的 Log 数据，我们怎么查找呢？可以按下 Ctrl+F 组合按键，在弹出的搜索框中输入 mtk_helloworld，然后点击 Find All 按钮，如图 2.5-32 所示。

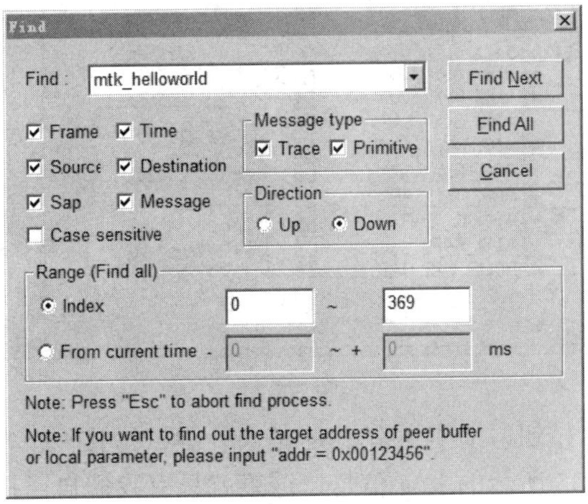

图 2.5-32

搜索结果如图 2.5-33 所示。这个 Log 信息的出现就证明 mtk_helloworld 函数已经执行了。那么我们能不能只打印自己的 Log,而不显示系统 log 呢?答案肯定是可以的。回到 mtk_helloworld 函数中,我们先来解读一下打印 log 的那条语句:

kal_prompt_trace(MOD_XDM, "-mtk-helloworld-%d--%s--", __LINE__, __FILE__);

Type	Index	Fram..	Time	Local Time	Source	Destination	Message
Trace	84		5307	17:27:07:174 2017/08/01	MOD_XDM		-mtk_helloworld-13--plutommi\MyStudyApp\MyStudyAppMain\Src\MyStudyAppMain.c--

图 2.5-33

kal_prompt_trace 是一个可变参数函数,函数体被封装在库里面,我们看不到,它的原型为 void kal_prompt_trace(module_type mod_id, const kal_char *fmt, ...)。第一个参数 mod_id 是一个枚举变量,它的取值可以是图 2.5-29 中 PS Module/Class 对话框中的所有复选框名称,通过这个名称可以实现 Log 信息的过滤,后面的传参方式与 C 语言中的 Printf 一模一样,但是格式符不能是%f,其中 __LINE__ 和 __FILE__ 是宏定义,前者输出当前代码所在的行数,后者输出当前代码的文件名。

再次单击"Filter"按钮,在弹出的对话框中单击"All Off",去掉 PS Module/Class 对话框中的所有复选框选项,然后选中最后一个 MOD_XDM,如图 2.5-34 所示,单击"确定"按钮。

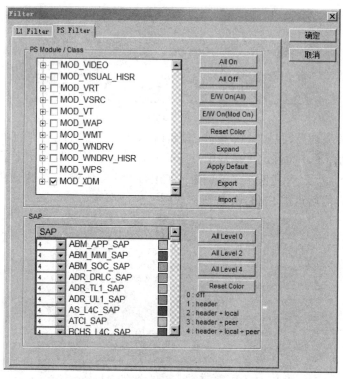

图 2.5-34

然后单击"Clear"按钮,清空所有 Log 信息。再次短按一下左软键(右上角的按键),此时就会出现如图 2.5-35 所示的 Log 信息,这里就只有我们自己打印的 Log,其他的 Log 都不再显示。

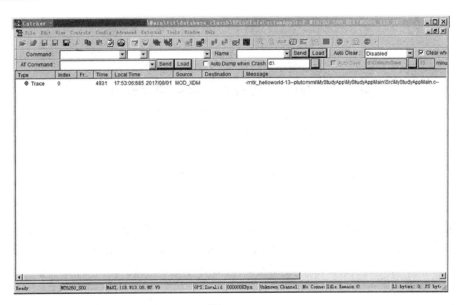

图 2.5-35

这里需要注意的是,选择 MOD id 的时候,尽量选择系统中使用比较少的,否则还是会出现其他的 Log 信息。笔者比较喜欢使用 MOD_XDM,这个 id 系统用得比较少,而且在 Filter 中很方便就能找到。读者可以选择自己喜欢的 MOD id。

第三章 MMI 基础编程

3.1 资　　源

大家都知道，手机是支持多国语言的，我们在国内买的手机，至少自带中文和英文两种语言，其中英文是必须要有的，因为 MTK 的 C 语言代码就是英文编写的，而其他语言包括中文，都可以根据项目需求自己配置。MTK 平台总共支持 80 种语言，基本上涵盖了全世界所有国家的语言。

在前面的示例程序中，在屏幕上显示的"hello world!"字符是在代码中直接输入的，但这只是英文版的，如果要显示中文版该如何处理呢？按照一贯的思维，肯定是分别定义不同语言版本的字符编码，然后用"if…erlse…"语句打印，这种方式并非不可以达到我们的目的，但试想一下，如果有 100 个字符串，要分别显示 80 种语言，这种方式是不是显得很笨拙？单"if…else…"的分支就有 8000 个了，虽然一个 MTK 设备不可能同时支持 80 种语言，但一个 MTK 设备中要显示的字符超过 100 个还是很正常的。所以这种方式行不通，而 MTK 系统也给我们提供了更好的解决方案——字符串资源。

MTK 中除了字符串资源外，还有图片、动画、铃声、屏幕、菜单、定时器、NVRAM、消息接收器等。所有的资源都以枚举 id 的形式存在和使用，它们在".res"文件中定义，使用 XML 语句编写，而".res"文件也称为资源文件。在使用 make resgen 命令编译的时候，会自动解析".res"文件中的资源，并生成对应的资源 id 头文件(.h)。

3.1.1 新增".res"资源文件

在 plutommi\MyStudyApp\MyStudyAppMain 目录下新建一个文件夹，命名为"Res"，然后在这个文件夹里面再新建一个文件，命名为"MyStudyAppMain.res"，并添加到 Source Insight 工程中(添加方法参照附录 B)。这个文件就作为我们的资源文件，目录结构如图 3.1-1 所示。

图 3.1-1

接下来我们就要开始编写资源文件。前面已经提到过，资源文件都是使用 XML 语句编写的，但是具体如何编写我们还是不知道，这里教给大家一个简单的学习方法，就是模仿 MTK 中原有的代码。随便打开一个 Res 文件，笔者找了一个内容比较简单的文件"Organizer.res"，这个文件的内容如图 3.1-2 所示。MTK 中所有的资源文件，不管其内容多么复杂，我们都可以把它们分为四个部分。

```
#include "mmi_features.h"
#include "custresdef.h"

/* Need this line to tell parser that XML start, must after all #include. */
<?xml version="1.0" encoding="UTF-8"?>

<APP id="APP_ORGANIZER">

    <INCLUDE file = "mmi_rp_app_mainmenu_def.h"/>

    <IMAGE id="ORGANIZER_TITLE_IMAGEID">CUST_IMG_PATH\\\\MainLCD\\\\Titlebar\\\\TB_OR.PNG</IMAGE>

    <SCREEN id="ORGANIZER_SCREENID"/>
    <SCREEN id="GRP_ID_ORG"/>

    <MENU id="MAIN_MENU_ORGANIZER_NO_SIM_MENUID" type="APP_SUB" str="MAIN_MENU_ORGANIZER_TEXT" img="ORGANIZER_TITLE_IMAGEID">
    </MENU>

</APP>
```

图 3.1-2

(1) #include 头文件部分。

这部分的内容跟 C 语言中包含的头文件是一模一样的，功能也是一模一样的。

(2) XML 文档序言：<?xml version = "1.0" encoding = "UTF-8"?>。

这部分内容只有一行代码。每个 XML 文档都是由 XML 序言开始，而 MTK 资源文件中的 XML 序言都是固定的，我们只需要照抄就行了。

(3) 定义资源 APP id：<APP id = "APP_ORGANIZER"> … </APP>。

APP_ORGANIZER 就是我们定义的资源 APP id，这个 id 的名称可以自己定义，通常使用大写，整个资源文件中的内容都必须包含在这两行代码之间。

(4) 资源定义部分。

在<APP id = "APP_ORGANIZER"> … </APP>之间的内容就是资源定义部分，其中包含了资源文件中能够定义的所有资源。

我们把这个文件的内容全部复制粘贴到 MyStudyAppMain.res 文件中，并用宏 __MYSTUDY_APPLICATION__ 包含起来，删除资源定义部分的内容，APP id 的名称 APP_ORGANIZER 更名为 APP_MYSTUDY，代码如下(代码清单 3.1-1)：

---代码清单 3.1-1---

#include "mmi_features.h"

#if defined(__MYSTUDY_APPLICATION__)
#include "custresdef.h"

/* Need this line to tell parser that XML start, must after all #include. */
<?xml version = "1.0" encoding = "UTF-8"?>

<APP id = "APP_MYSTUDY">

</APP>

```
#endif
```

一个资源文件的雏形已经写好了，但是我们要怎么让这个资源文件参与编译呢？我们定义的 APP_MYSTUDY 怎么使用呢？在前面添加 APP 的时候，MyStudyAppMain.mak 文件中只是添加了源文件，指定了头文件路径，并没有涉及资源文件。makefile 只是管理 C 语言文件(.c 和 .h 文件)，而 ".res" 文件并非 C 语言相关的文件，所以想要让 ".res" 文件参与编译，我们还需要在其他地方添加。我们照样从 MTK 的代码中学习，在 Source Insight 中全局搜索(用快捷键 Ctrl+/ 打开 Lookup References 窗口)APP_ORGANIZER，看看它在哪里被调用了。

在搜索的结果中，除了定义的资源文件外，还有很多文件中调用了 APP_ORGANIZER，但只有 Mmi_pluto_res_range_def.h 文件是有效的，其他文件都是编译过程中生成的(如果每次编译后，文件属性的修改时间都会变化，则该文件就是编译过程中生成的)，编译生成的文件，我们在其中修改的代码，最终也是无效的。所以这里的 APP id 应该在 plutommi\mmi\Inc\Mmi_pluto_res_range_def.h 文件中调用。打开这个文件，查看 APP_ORGANIZER 的调用语句如图 3.1-3 所示。

```
/***************************************************
*
****************************************************/
MMI_RES_DECLARE(APP_ORGANIZER, 50, ".\\mmi\\Organizer\\OrganizerRes\\")
/***************************************************
* CALENDAR application
****************************************************/
MMI_RES_DECLARE(APP_CALENDAR, 300, ".\\mmi\\Organizer\\OrganizerRes\\")
#define CALENDAR_BASE              ((U16) GET_RESOURCE_BASE(APP_CALENDAR))
#define CALENDAR_MAX               ((U16) GET_RESOURCE_MAX(APP_CALENDAR))
```

图 3.1-3

读者是否发现这个文件中 APP id 的使用方法有两种，关于 APP_ORGANIZER 的语句只有一行，但下面关于 APP_CALENDAR 的语句却有三行。这里 APP_ORGANIZER 的调用方法是以前老版本的方法，我们把这种方法舍弃，有兴趣的读者可以自己去看代码学习。APP_CALENDAR 使用新版本的方法添加资源更加简单，这也是笔者比较推荐的新增资源文件的方法，它总共只有三行代码。

(1) MMI_RES_DECLARE(APP_CALENDAR, 300, ".\\mmi\\Organizer\\OrganizerRes\\")

MMI_RES_DECLARE 是一个宏函数，它的作用是声明资源 APP id 中包含的资源最大个数(此处为 300)，并指定资源 APP id 所在资源文件的相对路径(此处为 ".\\mmi\\Organizer\\OrganizerRes\\")。指定了路径后，编译器就可以找到 ".res" 资源文件，并对它进行解析，让它参与系统编译。值得注意的是，这里的资源最大个数不是指所有资源个数相加的总和，而是各种资源分别可以定义的最大个数，比如字符串资源最多可以定义 300 个，图片资源最多也可以定义 300 个。

(2) #define CALENDAR_BASE ((U16) GET_RESOURCE_BASE(APP_CALENDAR))

CALENDAR_BASE 定义为资源 id 的最小值，GET_RESOURCE_BASE 同样也是宏函数。

(3) #define CALENDAR_MAX ((U16) GET_RESOURCE_MAX(APP_CALENDAR))

CALENDAR_MAX 定义为资源 id 的最大值,实际上 CALENDAR_MAX = CALENDAR_BASE+300。GET_RESOURCE_MAX 也是宏函数。

了解了以上三行代码后,我们依葫芦画瓢,把自己的资源 APP id 也添加到系统中。把上述三行代码全部复制粘贴到文件的最后一行 #endif 语句的前面,并用自己的宏包含起来,把 APP_CALENDAR 更名为 APP_MYSTUDY,资源文件的路径改为 ".\\MyStudyApp\\MyStudyAppMain\\Res\\",因为这里的资源 id 不需要 300 个,所以我们把资源个数改为 10,另外再把 CALENDAR_BASE 更名为 MYSTUDY_BASE,CALENDAR_MAX 更名为 MYSTUDY_MAX。Mmi_pluto_res_range_def.h 文件中添加代码如下(代码清单 3.1-2):

--代码清单 3.1-2--

```
/*省略部分代码*/
#ifdef __MMI_BT_NOTIFICATION__
MMI_RES_DECLARE(APP_BTNOTIFICATION, 50, ".\\MMI\\BTNotification\\BTNotificationRes\\")
#define BTNOTIFICATION_BASE ((U16) GET_RESOURCE_BASE(APP_BTNOTIFICATION))
#define BTNOTIFICATION_BASE_MAX ((U16) GET_RESOURCE_MAX(APP_BTNOTIFICATION))
#endif


#if defined(__MYSTUDY_APPLICATION__)
MMI_RES_DECLARE (APP_MYSTUDY, 10, ".\\MyStudyApp\\MyStudyAppMain\\Res\\")
#define MYSTUDY_BASE ((U16) GET_RESOURCE_BASE(APP_MYSTUDY))
#define MYSTUDY_MAX ((U16) GET_RESOURCE_MAX(APP_MYSTUDY))
#endif


#endif /* !defined(__COSMOS_MMI_PACKAGE__) */
```

到这里,资源文件就已经添加到系统中,接下来我们就可以使用 make resgen 指令编译资源了。编译成功后,会在 plutommi\Customer\CustomerInc 目录下生成一个名为 mmi_rp_app_mystudy_def.h 的文件,这个文件就是由资源 MyStudyAppMain.res 解析生成的,里面包含了我们定义的资源 id,每一类资源都会生成一个单独的枚举类型。如果我们需要使用这些资源 id,就必须包含这个头文件。

所有的资源文件编译成功后,都会有一个对应的资源 id 头文件。这个头文件的命名规则为 "mmi_rp_"+"小写 APP id"+"_def".h,比如我们定义的 APP id 为 APP_MYSTUDY,所以生成的头文件名为 "mmi_rp_app_mystudy_def.h",而 APP_CALENDAR 对应的资源 id 头文件为 "mmi_rp_app_calendar_def.h"。

3.1.2 字符串资源

上面定义的资源文件中,我们尚未在里面添加任何资源。这一节我们就把前面例子中的 "hello world!" 通过字符串资源的方式显示出来,以解决多国语言的显示问题。

MTK 中的字符资源全部添加在 plutommi\Customer\CustResource\PLUTO_MMI\ref_list.txt 文件中，它通过一个字符串资源 id 可以定义多个国家的不同语言，这个字符串资源 id 显示成何种语言会随系统的语言设置而变化。如果要新增对某种语言的支持，那我们只要对这个文件进行翻译就可以了，而不用修改代码。

MTK 提供了一个专用于修改字符资源的工具——plutommi\Customer\ STMTView.exe。我们双击打开它，然后点击 File→Open STMT... 会显示左右两个表格栏，随便选择一个，点击"open file"按钮，在弹出的文件选择对话框中，选择 plutommi\Customer\CustResource\PLUTO_MMI\ref_list.txt 然后打开，出现如图 3.1-4 所示的界面。

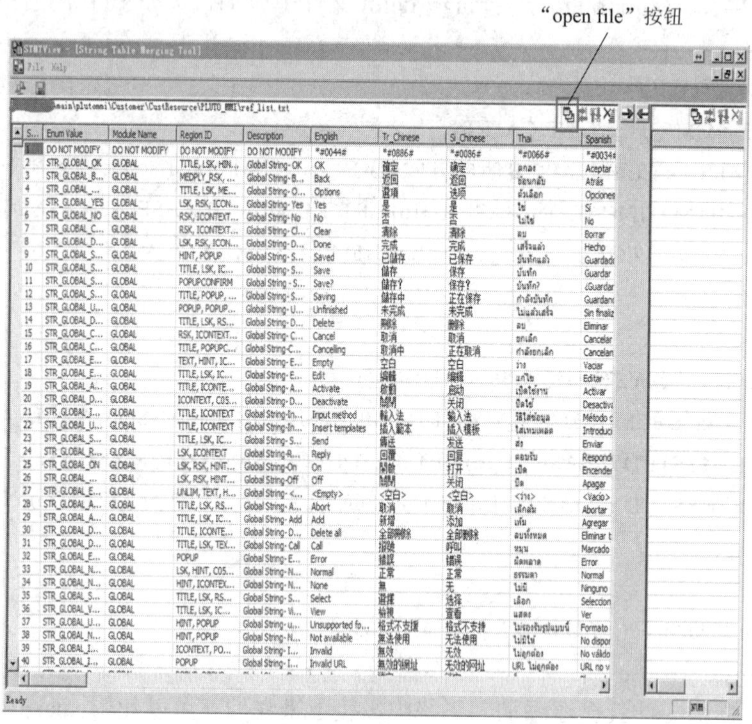

图 3.1-4

在打开文件的一栏中任意位置单击鼠标右键，选择"Insert new row"，在弹出的对话框中，新增一行字符串资源，如图 3.1-5 所示。

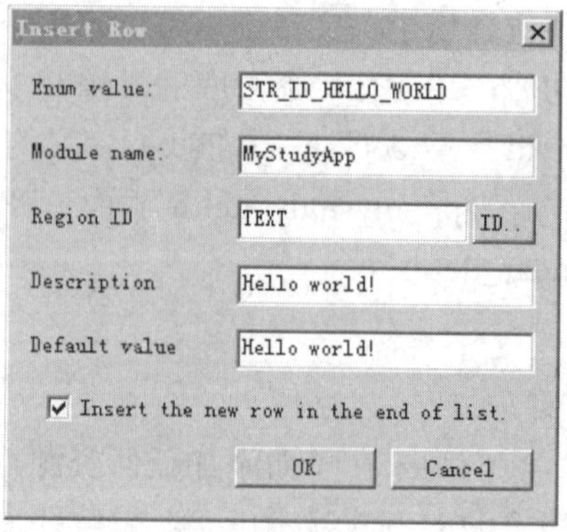

图 3.1-5

(1) Enum value：字符串资源 id，名称必须全部大写。这个 id 后面会在资源文件 MyStudyAppMain.res 中添加，否则编译资源的时候，依旧无法加载该字符串资源。

(2) Module name：字符资源所在的功能模块，这里可以使用默认的值，也可使用我们的模块名称，无特殊含义，仅作为描述。

(3) Region id：仅起描述作用，可以手动输入，也可以点击旁边的按钮选择，亦可不填。

(4) Description：描述字符 id 的用途。

(5) Default value：字符资源 id 的默认值，默认值通常为英文，在任何语言下都会默认使用这个值。

下面的复选框打钩，把资源 id 添加到文件的最后一行，点击"OK"按钮。在新增的一行对应 Si_Chinese 列把 "Hello world !" 改成 "你好，世界!"，添加效果如图 3.1-6 所示。

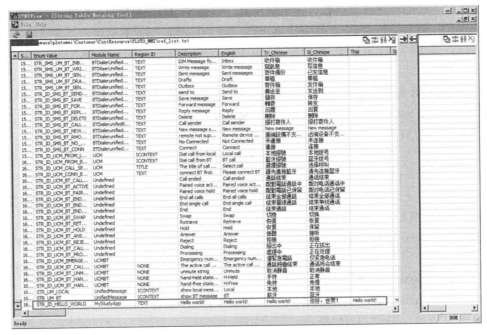

图 3.1-6

字符资源已经准备好了，接下来我们把新增的字符串资源 id "STR_ID_HELLO_WORLD" 定义到资源文件 MyStudyAppMain.res 中。添加语句如下(代码清单 3.1-3)：

---代码清单 3.1-3---

#include "mmi_features.h"

#if defined(__MYSTUDY_APPLICATION__)
#include "custresdef.h"

/* Need this line to tell parser that XML start, must after all #include. */
<?xml version = "1.0" encoding = "UTF-8"?>

```
<APP id = "APP_MYSTUDY">

    <!---------------符串资源-------------------------------------------------------------------->
        <STRING id = "STR_ID_HELLO_WORLD">"hello world!"</STRING>
</APP>
#endif
```

字符资源id命名通常以"STR_"开头，在资源文件中字符串资源的定义方式有两种：

```
<STRING id = "STR_NAME" > "default string" </STRING>
```

或

```
<STRING id = " STR_NAME"/>
```

这两种方法的区别在于前者设置了默认值，如果STR_NAME在ref_list.txt中不存在，那么它在任何语言下都会显示"default string"(注意：此处添加的默认值只能是英文字符)。而后者没有默认值，如果STR_NAME在ref_list.txt中不存在，则STR_NAME依旧是一个有效的资源id，此时它显示的内容我们无法预知，读者可以自己测试一下。

在ref_list.txt中我们添加了简体中文的字符，所以要让系统支持简体中文，否则我们无法看到效果。在Source Insight中打开 MMI_features_switchHEXING60A_11B.h文件，在其中找到宏开关CFG_MMI_LANG_SM_CHINESE，把后面的(_OFF_)改为(_ON_)。这个文件定义了整个源码中MMI层功能宏的开关。需注意的是，这个文件的名称及路径与项目名称是相关的，在目录plutommi\Customer\CustResource下有一个与项目名称一样的文件夹，该文件的命名规则就是"MMI_features_switch项目名称.h"，比如项目名称为"HEXING60A_11B_MMI"，则该文件名及其路径为"plutommi\Customer\CustResource\HEXING60A_11B_MMI\MMI_features_switchHEXING60A_11B.h"；若项目名称为"UMEOX60M_11B_MMI"，则文件名路径为"plutommi\Customer\CustResource\UMEOX60M_11B_MMI\MMI_features_switchUMEOX60M_11B.h"；若项目名称为"HEXING02A_WT_11C_MMI"，则该文件名及其路径为"plutommi\Customer\CustResource\HEXING02A_WT_11C_MMI\MMI_features_switchHEXING02A_WT_11C.h"。

另外，不同语言的编码方式是不一样的，所以还需要为系统添加字库。字库分为点阵字库和矢量字库，点阵字库存储在 vendor\font\FontData\OfficialFont 目录下对应的文件夹中，文件后缀名为bdf。MTK中自带的点阵字库通常只有英文字符和数字。如果需要支持其他语言，我们就需要另外新增字库，但点阵字库文件的获取途径非常有限，在智能穿戴的研发中，字库并不太重要，所以在此也不多作讲解。

矢量字库保存在plutommi\Customer\Fonts目录下对应的文件夹中，文件后缀名为ttf。如果平台不支持矢量字库(比如 6260M)，则系统中不会存在这个目录。在支持矢量字库的平台上，我们不需要考虑怎么获取字库文件，一个矢量字库文件可以支持多种语言，如果确实需要更换字库文件，则可以从网上下载或者从Windows系统中获取。虽然矢量字库比较占空间，但使用起来比较方便，如果空间允许，我们可以使用矢量字库，在HEXING60A_11B_GPRS.mak文件中设置 FONT_ENGINE = FONT_ENGINE_ETRUMP(儿童定位智能手表的代码上不能开这个宏)，然后使用make new指令编译一遍代码，再使用

make gen_modis 指令重新生成模拟器。

这里需要注意的是，如果使用的是儿童定位智能手表配套的 MTK6260M 代码，使用 make new 编译到最后会报错，因为打开中文宏开关 CFG_MMI_LANG_SM_CHINESE 后，系统存储空间不够，但不影响我们编译模拟器及使用模拟器。

编译完成后，在 Source Insight 工程中再次执行 Add tree，新增文件到项目中，然后全局搜索字符资源 id"STR_ID_HELLO_WORLD"。可以看到在 mmi_rp_app_mystudy_def.h 和 string_resource_usage.txt 文件中都包含了 STR_ID_HELLO_WORLD。

mmi_rp_app_mystudy_def.h 文件前面已经介绍过了，string_resource_usage.txt 文件在 plutommi\Customer\ResGenerator\debug\目录下，也是编译生成的，它包含了整个 MTK 系统的所有字符串资源，如果我们新增的字符串资源编译后能够在这个文件中找到，就说明添加成功了，否则就是新增资源失败。string_resource_usage.txt 中包含的字符串资源都可以直接在代码中使用，但是要包含对应的头文件。

在 MyStudyAppMain.c 文件中包含 mmi_rp_app_mystudy_def.h 头文件，并把 "gui_print_text(L"Hello world!");" 改为 "gui_print_text(GetString(STR_ID_HELLO_WORLD));"，代码如下(代码清单 3.1-4)：

---代码清单 3.1-4---

```
#if defined(__MYSTUDY_APPLICATION__)
#include "gui_themes.h"
#include "mmi_rp_app_mystudy_def.h"

void mtk_helloworld(void)
{
    gui_set_text_color(UI_COLOR_BLACK);        /*设置字符颜色为黑色*/
    gui_move_text_cursor(10, 10);              /*移动字符显示坐标为(10)*/
    gui_set_font(&MMI_large_font);             /*设置字符显示的字体*/

    gui_print_text((UI_string_type)GetString(STR_ID_HELLO_WORLD));/*在屏幕上打印字符*/

    kal_prompt_trace(MOD_XDM, "-mtk_helloworld-%d--%s--", __LINE__, __FILE__);

    gui_BLT_double_buffer(0, 0, UI_DEVICE_WIDTH, UI_DEVICE_HEIGHT);
}
#endif
```

运行模拟器，发现显示效果跟之前完全一样(编译 make r mystudyap 也许会报数据类型不匹配的错误，请读者自己修正，这是 C 语言的基础知识)。接下来我们把系统语言切换为中文，语言切换有一个比较快捷的方法，在 ref_list.txt 文件对应语言列的第一行，有类似 "*#XXXX#" 的字符编码，这些编码就是切换到对应语言的快捷指令，我们可以以拨号的方式使用。当然，如果对应语言不存在，那么其切换指令也不存在，此时使用指令切换语

言可走拨号的流程。

在 idle 界面直接按数字键盘输入"*#0086#"再按拨号键，就可以把系统语言切换到简体中文了，如图 3.1-7 所示。

图 3.1-7

语言切换之后，再次按 KEY_LSK(左软键)执行 mtk_helloworld 函数，此时显示为中文"你好，世界！"，如图 3.1-8 所示。

在使用字符串资源 id 的时候，我们使用了一个函数 GetString，这个函数会根据当前系统的语言环境，获取对应的字符串指针，而获取到的字符串默认就是 UCS2 编码，所以不需要进行编码转换。

MTK 中除了 C 语言中的一套字符串处理函数外，也有自己的一套处理函数。C 语言的字符串处理函数都是基于 ASC 编码，而 MTK 中的字符串处理函数基本上都是基于 UCS2 编码，它们大多定义在 UCS2.c 文件中，对应 C 语言中的字符串处理函数，比如字符串比较：strcmp、mmi_ucs2cmp；字符串拷贝函数 strcpy、mmi_ucs2cpy 等。另外，MTK 中关于字符串的显示函数大多以"gui_"开头命名，并以函数指针的形式存在，读者可查看"gui.h"文件；其次还有一些字符串编码转换的函数，读者可查看"Conversions.c"文件。

图 3.1-8

程序的功能是处理数据，而字符串是一个比较常见的数据表现形式。限于篇幅，这里不一一讲解，在后面的章节中，我们还会经常碰到字符串的处理，到时候再详细介绍。

3.1.3 屏幕资源

这里所说的屏幕并非指 MTK 设备上的物理硬件，而是指 MTK 系统中显示的界面。在 MTK 系统中，通常每一个界面都对应唯一一个屏幕 id，屏幕之间的相互切换及屏幕历史堆栈管理都是通过屏幕 id 进行的。屏幕资源的添加方法使用如下语句：

<SCREEN id = "SCREEN_NAME"/> 或者 <SCREEN id = "GROUP_NAME"/>

屏幕 id 资源的命名方式习惯上以 SCR_开头，比如<SCREEN id = "SCR_ID_IDLE_MAIN"/>，或者以 GRP_开头，比如<SCREEN id = "GRP_ID_IDLE_MAIN"/>。通常情况下，以 SCR_开头的资源一般作为实际的单个屏幕，而以 GRP_开头的资源都用于屏幕组(group)，一个 GRP 屏幕组下面，可以包含多个 SCR 屏幕，当你关闭一个 GRP 屏幕组时，其下的所有 SCR 屏幕就都被关闭了。在 MTK 智能设备开发中，GRP 屏幕组基本上用不到，SCR 屏幕 id 已经能够满足需求了，而且为了操作的简便也要尽量避免屏幕的频繁切换。实际上 SCR 和 GRP 都是同类型的资源，在使用的时候反过来也没有问题，但我们反对这么做，因为从命名上来看会导致歧义，减弱代码的可读性和可维护性。在 MTK 中所有的屏幕都呈一个树形结构，如图 3.1-9 所示。

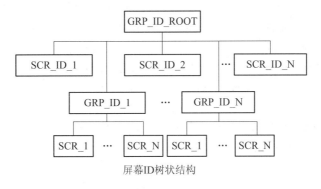

屏幕ID树状结构

图 3.1-9

接下来，我们给界面新增一个屏幕资源 id。在 MyStudyAppMain.res 文件中添加一行代码<SCREEN id = "SCR_ID_HELLO_WORLD"/>，代码如下(代码清单 3.1-5)：

--代码清单 3.1-5--
#include "mmi_features.h"

#if defined(_MYSTUDY_APPLICATION_)
#include "custresdef.h"

/* Need this line to tell parser that XML start, must after all #include. */
<?xml version = "1.0" encoding = "UTF-8"?>

<APP id = "APP_MYSTUDY">

    <!----------------符串资源-->
    <STRING id = "STR_ID_HELLO_WORLD" > "hello world!" </STRING>

    <!----------------屏幕资源-->
    <SCREEN id = "SCR_ID_HELLO_WORLD"/>

```
</APP>
#endif
```

然后用 make resgen 编译，编译完成之后，如果能够在 Mmi_rp_app_mystudy_def.h 文件中找到屏幕资源 id 即 SCR_ID_HELLO_WORLD，则说明屏幕资源添加成功了。

屏幕资源 id 添加成功后，我们把这个界面放到一个屏幕中管理，这里需要用到一个函数，我们称之为入屏函数，它的功能就是为当前的界面添加一个屏幕节点，其原型和参数说明如下：

```
MMI_BOOL    mmi_frm_scrn_enter (
    MMI_ID parent_id,           /*屏幕父节点，如果没有自己的屏幕组(GRP) id，可以使用
                                  根节点 GRP_ID_ROOT */
    MMI_ID scrn_id,             /*当前屏幕的 id*/
    FuncPtr exit_proc,          /*退出屏幕时的处理函数*/
    FuncPtr entry_proc,         /*进入屏幕时的处理函数*/
    mmi_frm_scrn_type_enum scrn_type    /*屏幕类型*/
)
```

通常在绘制界面之前，调用这个函数，并把绘制界面的函数指定为进入屏幕时的处理函数，MyStudyAppMain.c 文件中修改代码如下(代码清单 3.1-6)，省略了头文件包含部分，希望读者可以自己补充。

--代码清单 3.1-6--

```
/*省略部分代码*/
void mtk_helloworld_exit(void)
{
}
void mtk_helloworld(void)
{
    mmi_frm_scrn_enter(GRP_ID_ROOT, SCR_ID_HELLO_WORLD, mtk_helloworld_exit,
                       mtk_helloworld, MMI_FRM_FULL_SCRN);

    gui_set_text_color(UI_COLOR_BLACK);/*设置字符颜色为黑色*/
    gui_move_text_cursor(10, 10);/*移动字符显示坐标为(10)*/
    gui_set_font(&MMI_large_font);/*设置字符显示的字体*/

    gui_print_text((UI_string_type)GetString(STR_ID_HELLO_WORLD));/*在屏幕上打印字符*/

    kal_prompt_trace(MOD_XDM, "-mtk_helloworld-%d--%s--", __LINE__, __FILE__);

    gui_BLT_double_buffer(0, 0, UI_DEVICE_WIDTH, UI_DEVICE_HEIGHT);
}
```

再次运行模拟器，按下 KEY_LSK 键，其实显示效果没有任何变化，但是当我们按下 KEY_END 键时，屏幕上显示的字符就消失了，因为此时退出了自己的屏幕，回到了 idle 界面，系统会重新绘制 idle 界面的内容。而在添加屏幕之前，显示的字符实际上也是在 idle 界面上绘制的。为每一个界面添加一个屏幕，在 MTK 编程中是一个很好的习惯，这样做的好处很多，比如每进入一个新的屏幕时，按键的功能都可以重新定义，而且与其他屏幕中的按键功能不冲突；在退出屏幕的时候，可以在退出屏幕的处理函数中释放资源等。

这里再教大家一个小技巧，我们每进入一个屏幕，都要调用入屏函数 mmi_frm_scrn_enter，那么反过来我们只要跟踪这个函数的调用情况，就可以找到任何一个界面的绘制函数。这个方法在模拟器中通过打断点的方法跟踪很有效。

3.1.4 图片资源

MTK 中显示的图片有三种：一是通过资源加载的；二是显示磁盘中的文件；三是直接把图片转成二进制数组显示，这种方式常用于 SP(Service Provider，移动网信息服务业务) 游戏或应用开发。图片又分为静态图片和动态图片，静态图片就称为"图片"，而动态图片称为"动画"。实际上不管何种方式显示的何种图片，最终都是二进制数据，读者可查看 plutommi\Customer\CustResource 目录下类似于 "CustImg**.c" 之类的文件，比如 "CustImgDataRes_0.c"，这里面都是一些由资源生成的图片二进制数据。所有图片资源的添加方式都是一样的，语句格式为：

<IMAGE id="IMG_NAME">CUST_IMG_PATH"image_filepath\\\\image_filename"
</IMAGE>

图片资源 id 命名通常以 "IMG_" 开头，所有的图片资源都放在 plutommi\Customer\Image 文件夹中，在这个文件夹中包含不同屏幕尺寸的图片资源，每个屏幕尺寸对应的文件夹下都有一个 image.zip 压缩包，系统中的图片资源就放在这个压缩包中，我们添加的图片也要放到压缩包对应的目录中，在编译资源的时候，系统会将其解压成 MainLCD 目录。CUST_IMG_PATH 是一个宏，定义在 plutommi\Customer\CustResource\PLUTO_MMI\CustResDefPLUTO.h 文件中，它指向 image.zip 所在的目录，比如当前屏幕尺寸为 240×320(屏幕尺寸的配置在 HEXING60A_11B_GPRS.mak 文件中查看 MAIN_LCD_SIZE 宏定义)，则 CUST_IMG_PATH 定义指向 plutommi\Customer\Images\PLUTO240x320 目录，系统中定义的代码如下：

```
#define CUST_IMG_PATH                    "..\\\\..\\\\Customer\\\\Images\\\\PLUTO176X220"
#elif defined(__MMI_MAINLCD_240X320__)
    #define CUST_IMG_PATH                "..\\\\..\\\\Customer\\\\Images\\\\PLUTO240X320"
#elif defined(__MMI_MAINLCD_320X240__)
    #define CUST_IMG_PATH                "..\\\\..\\\\Customer\\\\Images\\\\PLUTO320X240"
```

image_filepath 是相对于 MainLCD 的目录，image_filename 就是图片的文件名，包含后缀名。图片资源的格式可以是 bmp、png、jpg、gif 以及 MTK 系统特定的一些图片格式。系统中添加成功的图片资源，其资源 id 都可以在 plutommi\Customer\ResGenerator\debug\image_resource_usage.txt 文件中找到。如果资源加载失败，则资源 id 会被添加到相同目录的 image_load_fail.txt 文件中，这种情况通常都是图片的目录错误，或者 image.zip 中没有这个图片文件。

进入目录 plutommi\Customer\Images\PLUTO240x320，我们要在 image.zip 压缩包中新建一个自己的图片资源目录，为了方便，我们先添加到 MainLCD 中，验证成功后再合并到 image.zip 压缩包中。为了避免编译资源时会解压 image.zip 替换掉 MainLCD，我们要在 mtk_resgenerator.cpp 文件的 main 函数中把 UnzipImage();语句先注释掉，如图 3.1-10 所示。

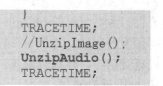

图 3.1-10

在 MainLCD 目录下新建一个文件夹，命名为 MyStudyApp，作为我们自己模块的图片资源目录，读者可随便找两个图片文件比如一个静态图片 "image_static.png" 和一个动态图片 "image_anim.gif" 放进去，但图片的尺寸不要太大，否则屏幕上显示不完，笔者的图片尺寸定为 60×60，如图 3.1-11 所示。

图 3.1-11

然后在 MyStudyAppMain.res 文件中添加两行代码，加载图片资源。代码如下(代码清单 3.1-7)：

--代码清单 3.1-7--

```
#include "mmi_features.h"

#if defined(__MYSTUDY_APPLICATION__)
#include "custresdef.h"

/* Need this line to tell parser that XML start, must after all #include. */
<?xml version="1.0" encoding="UTF-8"?>

<APP id="APP_MYSTUDY">
    <!----------------符串资源---------------------------------------------------------------->
    <STRING id="STR_ID_HELLO_WORLD">"hello world!"</STRING>
```

```
<!---------------图片资源-------------------------------------------------------->
<IMAGE id="IMG_STATIC_ID">CUST_IMG_PATH"\\\\MainLCD\\\\MyStudyApp\\\\image_static.png"</IMAGE>
<IMAGE id="IMG_ANIM_ID">CUST_IMG_PATH"\\\\MainLCD\\\\MyStudyApp\\\\image_anim.gif"</IMAGE>

<!---------------屏幕资源-------------------------------------->
<SCREEN id="SCR_ID_HELLO_WORLD"/>

</APP>
#endif
```

修改资源都必须使用 make resgen 编译，编译完成之后在 image_resource_usage.txt 看是否能找到我们新增的两个图片资源 id，如果能找到，说明资源加载成功。回到 mtk_resgenerator.cpp 文件，去掉 main 函数中 UnzipImage();语句的注释，把 plutommi\Customer\Images\PLUTO240x320 目录下 MainLCD 中添加的 MyStudyApp 目录合并到 image.zip 压缩包的 MainLCD 目录中，删除已解压出来的 MainLCD 文件夹，再次编译资源，此时会重新解压 image.zip，生成 MainLCD 目录，并且新增的图片文件也存在。

关于动画资源，在 MTK 系统中还有另外一种存在形式，比如插入充电器时，状态栏上的电池充电动画，资源路径为 MainLCD\IdleScreen\Statusicons\battery\SI_BAT 目录，如图 3.1-12 所示。

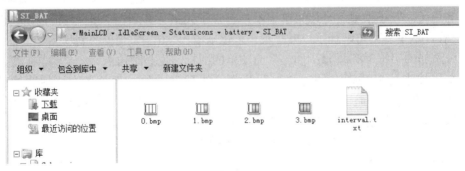

图 3.1-12

其中名称为 0～3 的 bmp 图片就是动画的第 0～3 帧，而文件 interval.txt 中描述的是每两帧之间的延时，加载资源时只需要指向 SI_BAT 目录即可，不需要文件后缀名。比如：

```
<IMAGE id = "IMG_ANIM">
CUST_IMG_PATH\\\\MainLCD\\\\IdleScreen\\\\StatusIcons\\\\battery\\\\SI_BAT
</IMAGE>;
```

这种形式的动画资源与 GIF 动画资源没有什么区别，在使用的时候也是一样的。关于图片资源的显示方式，我们放到后面的章节中去讲解。

3.1.5 菜单资源

在讲解菜单之前，先认识一下 MTK 中的菜单。把 mainmenu.c 文件中自己添加的代码全部去掉，运行模拟器，按下 KEY_LSK 键进入主菜单界面，这个界面显示的是矩阵菜单。再按一下 KEY_LSK 键进入短信界面，这个界面显示的是列表菜单，如图 3.1-13 所示。

图 3.1-13

每个菜单都有自己的高亮函数，就是我们选择这个菜单时所执行的函数，在高亮函数里面通常只是注册按键功能。如图 3.1-13 所示的"信息"菜单被选中后，按下左软键就会进入右侧的信息菜单列表界面，这个功能就是在信息菜单的高亮函数里面实现的。此外菜单还有自己的名字、图片等属性。菜单与屏幕一样，在系统中也是呈树状结构，通常称之为菜单树，根节点是 IDLE_SCREEN_MENU_ID。菜单资源的添加方法稍微复杂些，其格式主要有以下三种：

(1) 叶子菜单，代码如下：

<MENUITEM id="MENU_ID"str="MENU_NAME" img="MENU_ICON"highlight="highlight_func" />

顾名思义，叶子菜单就是该菜单下不再包含子菜单。其属性"id"与前面所讲解的资源 id 一样，这里就不再解释了。叶子菜单可以缺省；"str"为菜单的名称，这里必须传入一个已经添加的字符串资源 id；"img"为菜单的图标，也必须传入一个已添加的图片资源 id，这个属性也可以缺省；"highlight"是菜单的高亮函数。另外还有快捷方式的属性"shortcut"和"shortcut_img"、"launch"，这三个属性在 MTK 智能设备研发中基本上用不到，所以本书不作讲解，有兴趣的读者可以通过 MTK 中自带的源码学习。还有菜单类型"type"属性，目前没有发现有什么用处，我们可以不用它。

(2) 根节点菜单，代码如下：

<MENU id="MENU_ID" str="MENU_NAME" img="MENU_ICON" highlight="highlight_func" >
    <MENUITEM_ID>MENU_ITEM1_ID</MENUITEM_ID>
    <MENUITEM_ID>MENU_ITEM2_ID</MENUITEM_ID>
</MENU>

以上 MENU_ITEM1_ID 和 MENU_ITEM2_ID 是已经添加好的叶子菜单，如果还没有添加，也可以写成下面这种方式(省略了其他属性)，直接在指定子菜单的时候添加。

<MENU id="MENU_ID" str="MENU_NAME" img="MENU_ICON" highlight="highlight_func" >

```
    <MENUITEM id="MENU_ITEM1_ID" str="MENU_ITEM1_NAME" img="MENU_ITEM1_ICON"/>
    <MENUITEM id="MENU_ITEM2_ID" str="MENU_ITEM2_NAME" img="MENU_ITEM2_ICON"/>
</MENU>
```

根节点菜单也可以作为其他根节点的子菜单,当然根节点菜单也可以不包含任何子菜单。

(3) 主菜单,代码如下:

```
<MAINMENUITEM id="MAIN_MENU_ID"
    #ifdef __MMI_MAINMENU_LIST_SUPPORT__
        mm_list_img="MAIN_MENU_LIST_ICON"
    #endif
    #ifdef __MMI_MAINMENU_MATRIX_SUPPORT__
        mm_matrix_img="MAIN_MENU_MATRIX_ICON"
        mm_matrix_highlight_img="MAIN_MENU_MATRIX_HIGHLIGHT_ICON"
    #endif
    #ifdef __MMI_MAINMENU_PAGE_SUPPORT__
        mm_page_img="MAIN_MENU_PAGE_ICON"
    #endif
        mm_title_img="MAIN_MENU_TITLE_ICON"
/>
```

系统中的主菜单基本上都添加在"MainMenuRes.res"文件中,它有三种形式,即列表、页面和矩阵,对应的属性都分别用宏控制起来,其中属性"id"必须是已经存在的菜单资源 id,其他的"img"也必须是已经存在的图片资源 id。矩阵类型的主菜单有两个图片属性,其中"mm_matrix_img"是未选中时的菜单 icon,"mm_matrix_highlight_img"是选中时的菜单 icon,这两个 icon 也可以用同一个图片资源 id。

菜单资源编译成功后,都可以在 plutommi\Customer\CustResource\ CustMenuTree_Out.c 文件中找到,并且这个文件还列出了系统中所有菜单的树状结构关系。如果定义菜单资源 id 时,指定了"highlight"属性,则编译资源后在 plutommi\Framework\EventHandling\EventsInc\mmi_menu_handlers.h 文件的 mmi_frm_const_hilite_hdlr_table 结构体变量中会生成菜单 id 与高亮函数的对应关系。

接下来我们就给自己的代码添加菜单入口。在 MyStudyAppMain.res 文件中添加代码如下(代码清单 3.1-8):

```
------------------------------------MyStudyAppMain.res 代码清单 3.1-8--------------------------------
#include "mmi_features.h"


#if defined(__MYSTUDY_APPLICATION__)
#include "custresdef.h"


/* Need this line to tell parser that XML start, must after all #include. */
<?xml version="1.0" encoding="UTF-8"?>
```

```
<APP id="APP_MYSTUDY">
    <!----------------Include Area---------------------------------------------------------------->
    <INCLUDE file="GlobalResDef.h, CustMenuRes.h, GlobalMenuItems.h"/>

    <!----------------符串资源--------------------------------------------------------------------->
    <STRING id="STR_ID_HELLO_WORLD">"hello world!"</STRING>

    <!----------------图片资源--------------------------------------------------------------------->
    <IMAGE id="IMG_STATIC_ID">CUST_IMG_PATH"
     \\\\MainLCD\\\\MyStudyApp\\\\image_static.png"</IMAGE>
    <IMAGE id="IMG_ANIM_ID">CUST_IMG_PATH"
     \\\\MainLCD\\\\MyStudyApp\\\\image_anim.gif"</IMAGE>

    <!----------------菜单资源--------------------------------------------------------------------->
    <MENUITEM id="MENU_HELLO_WORLD_ID" str="STR_ID_HELLO_WORLD"
              highlight="mmi_highlight_helloworld"/>

    <MENU id="MAIN_MENU_MYSTUDY_ID" str="STR_ID_HELLO_WORLD"
              img="IMG_STATIC_ID" highlight="mmi_highlight_mystudy">
        <MENUITEM_ID>MENU_HELLO_WORLD_ID</MENUITEM_ID>
    </MENU>

    <MAINMENUITEM id="MAIN_MENU_MYSTUDY_ID"
    #ifdef __MMI_MAINMENU_MATRIX_SUPPORT__
        mm_matrix_img="IMG_STATIC_ID"
        mm_matrix_highlight_img="IMG_ANIM_ID"
    #endif
        mm_title_img="IMG_STATIC_ID"
    />

    <!----------------屏幕资源--------------------------------------------------------------------->
    <SCREEN id="SCR_ID_HELLO_WORLD"/>

</APP>
#endif
```

代码中，我们添加了两个菜单。一个是叶子菜单 MENU_HELLO_WORLD_ID，菜单的名称是我们前面新增的字符串资源 id 即 ISTR_ID_HELLO_WORLD，菜单 icon 缺省，高亮函数是 mmi_highlight_helloworld，这个函数之前并没有，所以我们还要在 MyStudyAppMain.c

文件中定义它，它的函数原型为 void mmi_highlight_helloworld(void)，函数体暂时为空，什么也不做。另一个菜单是根节点菜单 MAIN_MENU_MYSTUDY_ID，菜单名字也是 STR_ID_HELLO_WORLD 字符串资源 ID，菜单 icon 为 IMG_STATIC_ID，高亮函数为 mmi_highlight_mystudy，也需要自己定义，函数原型为 void mmi_highlight_mystudy(void)，函数体暂时也置为空。它包含子菜单 MENU_HELLO_WORLD_ID。我们把这个菜单作为主菜单，主菜单未选中状态下的 icon 和标题 icon 都用静态图片 IMG_STATIC_ID，选中状态下的 icon 用动画 IMG_ANIM_ID。既然是主菜单，那我们就要把它挂在主菜单树中，在 MainMenuRes.res 文件中找到整个系统菜单树的根菜单 IDLE_SCREEN_MENU_ID，把 MAIN_MENU_MYSTUDY_ID 添加为它的子菜单，代码如下(代码清单 3.1-9)：

--代码清单 3.1-9--

```
/*省略默认代码*/
#elif  defined(_MMI_MAINLCD_176X220_) || defined(_MMI_MAINLCD_240X320_)  /*when matrix is 3X3 */
        <MENU id="IDLE_SCREEN_MENU_ID" type="MATRIX_MENU" str
            ="MAIN_MENU_MENU_TEXT" img="MAIN_MENU_PHONEBOOK_ICON">
/*省略默认代码*/
#if defined(_MMI_BT_DIALER_SUPPORT_) && !defined(_MMI_MRE_MAIN_MENU_)

        <MENUITEM_ID>MAIN_MENU_TOOLSANDFUN_MENUID</MENUITEM_ID>
#else
        <MENUITEM_ID>MAIN_MENU_TOOLS_MENUID</MENUITEM_ID>
#endif
#if defined(_MYSTUDY_APPLICATION_)
        <MENUITEM_ID>MAIN_MENU_MYSTUDY_ID</MENUITEM_ID>
#endif
        </MENU> /*defined(_MMI_MAINLCD_176X220_) ||
            defined(_MMI_MAINLCD_240X320_)*/
#else /* _MMI_MAINLCD_128X160_, _MMI_MAINLCD_240X400_, _MMI_MAINLCD_
    320X480_   */ /* when matrix is 3X4 */
/*省略默认代码*/
```

需要注意的是，IDLE_SCREEN_MENU_ID 在 MainMenuRes.res 文件的定义存在多个，但都用宏控制起来了，最终生效的只有一个，我们需要区分哪些宏是开着的，哪些宏是关着的。这里教大家一个简单的方法，一般命名中有 MMI 字符的宏，都定义在 plutommi\mmi\Inc\MMI_features.h 文件中，这些宏都可以在 build\HEXING60A_11B\log\MMI_features.log 文件中查看，只要能在这个文件中找到，并且宏的前面有字符"[D]"，则表示该宏已经定义了，如果前面是"[U]"，则说明没定义。定义在 Option.mak 文件和项目配置文件 HEXING60A_11B_GPRS.mak 中的宏，如果是有效的，都可以在 build\HEXING60A_11B\log\infomake.log 文件中找到；反过来说，只要在 infomake.log 文件中能够查找到的宏，都

是已经定义了的有效宏。

以上代码添加完后，执行 make resgen 编译资源，然后运行模拟器，进入主菜单，选中的状态如图 3.1-14 左所示，此时的静态图片就是 mm_matrix_img 属性设定的；选中状态下如图 3.1-14 右所示，此时的动画就是 mm_matrix_highlight_img 属性设定的，标题栏左侧显示 icon，是 mm_title_img 属性，标题栏显示的字符是 MAIN_MENU_MYSTUDY_ID 菜单的 str 属性。

图 3.1-14

虽然菜单显示出来了，但菜单对应的高亮处理函数还是空的，按下左软键时依旧进入的是上一个高亮菜单的功能。我们在 mmi_highlight_mystudy 函数里面注册左软键功能进入 mtk_helloworld 函数。代码如下(代码清单 3.1-10)：

---代码清单 3.1-10---
```
/*省略部分代码*/
void mmi_highlight_helloworld(void)
{
        SetKeyHandler(mtk_helloworld, KEY_LSK, KEY_EVENT_UP);
}
```

再次运行模拟器，选中自己新增的菜单，按下左软键，就会执行 mtk_helloworld 函数(读者可设置断点跟踪代码执行情况)，屏幕左上角会打印出"你好，世界！"，如图 3.1-15 所示。

如果代码就这样写，那我们前面还有一个叶子菜单就没用了，所以应该是在叶子菜单的高亮函数里面，注册左软键功能进入 mtk_helloworld，把 mmi_highlight_mystudy 函数中的语句移到 mmi_highlight_helloworld 函数中，另外新增一个函数 mmi_entry_mystudy 并在 mmi_highlight_mystudy 函数中注册为左软键事件。mmi_entry_mystudy 函数中的功能用于显示 MAIN_MENU_MYSTUDY_ID 菜单的子菜单列表。子菜单列表也是一个界面，我们再为它创

图 3.1-15

建一个屏幕 id：SCR_ID_MYSTUDY_MENU_LIST。MyStudyAppMain.c 文件的代码清单如下(代码清单 3.1-11)：

--MyStudyAppMain.c 代码清单 3.1-11--------------------------------

```c
#if defined(__MYSTUDY_APPLICATION__)
#include "gui_themes.h"
#include "mmi_rp_app_mystudy_def.h"
#include "Mmi_frm_scenario_gprot.h"
#include "GlobalResDef.h"

void mtk_helloworld_exit(void)
{
}
void mtk_helloworld(void)
{
    mmi_frm_scrn_enter(GRP_ID_ROOT, SCR_ID_HELLO_WORLD, mtk_helloworld_exit,
                    mtk_helloworld, MMI_FRM_FULL_SCRN);

    gui_set_text_color(UI_COLOR_BLACK);          /*设置字符颜色为黑色*/
    gui_move_text_cursor(10, 10);                /*移动字符显示坐标为(10)*/
    gui_set_font(&MMI_large_font);               /*设置字符显示的字体*/

    gui_print_text(GetString(STR_ID_HELLO_WORLD)) ;/*在屏幕上打印字符*/

    kal_prompt_trace(MOD_XDM, "-mtk_helloworld-%d--%s--", __LINE__, __FILE__);

    gui_BLT_double_buffer(0, 0, UI_DEVICE_WIDTH, UI_DEVICE_HEIGHT);
}

void mmi_highlight_helloworld(void)
{
    SetKeyHandler(mtk_helloworld, KEY_LSK, KEY_EVENT_UP);
}

void mmi_entry_mystudy(void)
{
    /*----------------------------------------------------------------*/
    /* Local Variables                                                */
    /*----------------------------------------------------------------*/
    U8 i = 0, *item_string[MAX_SUB_MENUS]={0x00};
```

```
U16 item_list[MAX_SUB_MENUS]={0x00}, item_icon[MAX_SUB_MENUS]={0x00},
                    num_of_items=0;

/*----------------------------------------------------------------*/
/* Code Body                                                      */
/*----------------------------------------------------------------*/

    mmi_frm_scrn_enter(GRP_ID_ROOT,     SCR_ID_MYSTUDY_MENU_LIST,     NULL,
mmi_entry_mystudy, MMI_FRM_FULL_SCRN);

    SetParentHandler(MAIN_MENU_MYSTUDY_ID);/*设置父类菜单 ID*/
    num_of_items = GetNumOfChild_Ext(MAIN_MENU_MYSTUDY_ID);/*获取子菜单的个数*/
    GetSequenceStringIds_Ext(MAIN_MENU_MYSTUDY_ID, item_list);
                                            /*获取子菜单名称字符串列表*/
    GetSequenceImageIds_Ext(MAIN_MENU_MYSTUDY_ID, item_icon);
                                            /*获取子菜单图标 icon 列表*/
    for (i = 0; i < num_of_items; i++)
    {
        item_string[i] = (U8*) GetString(item_list[i]);
    }

    /*执行当前选中菜单的高亮函数*/
    RegisterHighlightHandler(ExecuteCurrHiliteHandler);

    /*显示子菜单列表*/
    ShowCategory53Screen(
            STR_ID_HELLO_WORLD,
            0,
            STR_GLOBAL_OK,
            IMG_GLOBAL_OK,
            STR_GLOBAL_BACK,
            IMG_GLOBAL_BACK,
            num_of_items,
            item_string,
            item_icon,
            NULL,
            0,
            0,
            NULL);

    SetRightSoftkeyFunction(mmi_frm_scrn_close_active_id, KEY_EVENT_UP);
```

}

void mmi_highlight_mystudy(void)
{
    SetKeyHandler(mmi_entry_mystudy, KEY_LSK, KEY_EVENT_UP);
}
#endif

运行模拟器，当我们选中主菜单 MAIN_MENU_MYSTUDY_ID 后，进入如图 3.1-16 左图所示界面，这个界面以列表的形式显示了子菜单 MENU_HELLO_WORLD_ID，因为只有一个子菜单，所以默认就是选中状态，再次按下左软键，进入图 3.1-16 右图所示界面，屏幕上打印出"你好，世界！"。

图 3.1-16

在 mmi_entry_mystudy 中，用到了 MTK 中的一些函数，这些函数都是系统定义的接口，代码中都有注释，也可以找到它们的实现源码，希望读者能自己理解它们的调用方法。在 MTK 中还有很多类似于 ShowCategory53Screen 的函数接口，这类函数被称为"屏幕模板"，每个函数都封装了一个特定的屏幕显示效果，而且命名都类似于 ShowCategoryXXXScreen。MTK 的官方有专门展示屏幕模板显示效果的文档，但笔者希望读者可以自己从代码中学习，只要找到某个界面的入口函数，就可以模仿它的代码，比如图 3.1-16 所示的这个子菜单列表界面，笔者就是模仿工具箱菜单 MAIN_MENU_ORGANIZER_MENUID 的界面绘制函数 EntryOrganizer 而写的。

3.1.6 铃声资源

系统中播放的铃声与图片类似，也有三种形式：二进制数组、文件和铃声资源。铃声二进制数组主要用于游戏中；铃声文件通常用于音乐播放器，其他铃声比如按键音、来电铃声、闹钟铃声等通常都是用铃声资源。铃声资源添加方式也与图片资源相似，语句格式为：

    <AUDIO id = "AUD_ID_NAME" flag = "MULTIBIN">CUST_ADO_PATH"
        \\\\file_path \\\\file_name"</AUDIO>

"id"属性就不再介绍了，每个资源都有一个 id；"flag"属性目前也没发现有什么用处，可写可不写。CUST_ADO_PATH 是一个文件路径宏定义，指向 plutommi\Customer\Audio\PLUTO 目录，该目录下有一个压缩包 audio.zip，系统中所有的铃声文件都包含在这个压缩包中。编译资源的时候 audio.zip 会被解压到当前文件，所以我们添加的铃声文件也要放到这个压缩包中。"file_path"是 audio.zip 压缩包中的路径；"file_name"是铃声资源文件名，其后缀名主要有 mp3、wav、mid、imy 等。

系统中大多数铃声资源都在情景模式中，参见 ProfilesSrv.res 文件。铃声资源编译成功后，与图片资源类似，可以在 plutommi\Customer\ResGenerator\debug\audio_resource_usage.txt 文件中找到 id 以及与其对应的铃声文件，并且会在 plutommi\Customer\CustResource 目录下生成一些命名类似于"CustAdo*"的文件(比如"CustAdoDataHWExt.h")，这些文件中保存着铃声资源生成的二进制数据数组。

下面我们就来添加一个铃声资源。把 plutommi\Customer\Audio\PLUTO 目录下的所有文件和文件夹全部删除，只留下 audio.zip 压缩包文件，然后打开它，在里面新建一个文件夹，命名为"MyStudyApp"，用于保存我们自己模块的铃声资源，随便找一首自己喜欢的 mp3 音乐放进去，重命名为"music.mp3"，如图 3.1-17 所示。

图 3.1-17

在 MyStudyAppMain.res 文件中添加如下代码，加载铃声资源：

```
<AUDIO id="AUD_MY_MUSIC_ID" flag = "MULTIBIN">CUST_ADO_PATH
    "\\\\MyStudyApp\\\\music.mp3"</AUDIO>
```

最后一步，make resgen 编译资源，至于是否添加成功，就留给读者自己去判断。关于如何播放的问题，放到后面的章节中讲解。

3.1.7 NVRAM 资源

MTK 设备有些数据在关机之后也要保存，实现这种功能的方法有两种，一种方法是存储到磁盘文件中，另一种方法就是我们本节要讲的 NVRAM(简称 NV)。NVRAM 分为两种：一种可以存储复合数据类型，比如结构体；另一种只能存储数值类型，比如 byte、short、double，我们用得最多的是 byte 和 short。存储数值类型的 NV 就是通过资源的方式添加的，添加 NV 资源的格式为：

```
<CACHEDATA type = "type" id = "NVRAM_NAME" restore_flag="TRUE/FALSE">
    <DEFAULT_VALUE> [value] </DEFAULT_VALUE>
```

```
<DESCRIPTION> description </DESCRIPTION>
    <FIELD min = "min_number" max = "max_number"></FIELD>
</CACHEDATA>
```

其中：属性"type"为 NV 中存储的数据类型，其取值有 byte(1 个字节)、short(2 个字节)、double(8 个字节)；属性"restore_flag"标记恢复出厂设置的时候，该 NV 中存储的数据是否恢复为默认值；<DEFAULT_VALUE> [value] </DEFAULT_VALUE>字段中的 value 为 NV 的默认值，如果 type 为 byte，取值最大为[0xFF]；short 最大为[0xFF, 0xFF]，左侧为高位，右侧为低位；double 最大为[0xFF, 0xFF, 0xFF, 0xFF, 0xFF, 0xFF, 0xFF, 0xFF]，同样是左侧高位，右侧低位。比如 type 为 short，存储 value 为 300，转换成十六进制为[0x01, 0x2C]。

<DESCRIPTION>description</DESCRIPTION>字段是 NV 的描述，相当于注释，这个属性可以缺省；<FIELD min = "min_number" max = "max_number"></FIELD>字段描述 NV 的取值范围，也相当于注释的作用，可以缺省。比如，系统中保存背光亮度和背光时间的 NV，在 gpiosrv.res 文件中定义如下：

```
<CACGEDATA type="byte" id="NVRAM_BYTE_BL_SETTING_LEVEL" restore_flag="TRUE">
    <DEFAULT_VALUE> [0xFF] </DEFAULT_VALUE>
    <DESCRIPTION> Backlight Level </DESCRIPTION>
    <FIELD min="1" max="20"></FIELD>
</CACHEDATA>

<CACHEDATA type="short" id="NVRAM_BYTE_BL_SETTING_HFTIME" restore_flag="TRUE">
    <DEFAULT_VALUE> [0Xff, 0xFF] </DEFAULT_VALUE>
    <DESCRIPTION> Backlight timer </DESCRIPTION>
<CACHEDATA>
```

通过前面的学习，相信现在读者已经可以自己写一些代码了。这里就给读者布置一个作业，在 MyStudyAppMain.res 文件中添加一个 NV 资源 id，名称为"NVRAM_MYSTUDY_ID"，类型为"byte"，默认值为"0x00"，NV 资源编译后，如果在 custom\common\custom_mmi_cache_cat.xml 文件中有生成自己新增的 NV 资源 id，则说明添加成功了。NV 资源的使用方法在后面的章节讲解。

3.1.8 定时器资源

定时器的作用就是延后执行某一个操作，也可以每隔一段时间就执行一次。定时器 id 的添加方式有两种：一种是在 TimerEvents.h 文件中的 MMI_TIMER_IDS 枚举中添加，只要添加在 KEY_TIMER_ID_NONE 和 MAX_TIMERS 之间就可以了，这种添加方式比较简单，通常驱动里面要用的定时器都添加在这个文件中；另一种方式是通过资源 id 方式添加，这种方式添加的定时器跟第一种方式添加的定时器没有本质的区别，在使用上也是完全一模一样的。MMI 层用到的定时器两种添加方式都没有问题，但是强烈建议使用第二种，以方便代码的模块化。添加定时器资源的语句格式为：

```
<TIMER id = "TIMER_ID_NAME"/>
```

定时器资源的添加方式应该算是最简单的了，因为它只有一个资源 id 属性。在 MyStudyAppMain.res 文件中定义一个定时器资源 MYSTUDY_TIMER_ID，也交给读者自己完成。编译资源后，如果能在 mmi_rp_app_mystudy_def.h 文件中生成就算添加成功了。至于定时器的使用方法，放到后面的章节中讲解。

3.1.9 消息资源

通常在中断处理函数中无法做一些比较耗时的事情，但是当中断触发时，我们有时又不得不处理一些比较耗时的事情，这个时候就要用到消息。当中断触发时，在中断处理函数中发送消息，然后在消息接收函数中处理比较耗时的事情，这种处理机制有点类似于新建一个线程。定义消息资源需要一个消息发送者"SENDER"和一个消息接受者"RECEIVER"，语句格式如下：

```
<SENDER    id="MSG_ID_NAME "  hfile="headfile"/>
<RECEIVER  id="MSG_ID_NAME"   proc="reveiver_func"/>
```

"SENDER"为消息的发送者，"RECEIVER"为消息的接收者，它们两个的资源 id 必须同名，"RECEIVER"中的"proc"属性就是消息接收处理函数，其函数原型为 mmi_ret (*msg_recv_proc)(mmi_event_struct * evt)。值得注意的是，一个消息 id 可以有多个消息处理函数，也就是说可以定义多个 id 相同但"proc"不同的"RECEIVER"语句。同一个函数，也可以同时作为多个消息 id 的处理函数，但为了代码的可读性和可维护性，建议使用一对一的定义关系。

消息资源的 id 与其他资源 id 不一样，它无法编译生成到对应的资源头文件中，需要我们自己定义在属性"hfile"指定的头文件中，定义的方式与生成的资源头文件中其他资源 id 枚举类似，最小值也是从资源 APP id 的 BASE id 开始。"hfile"指定的头文件中还定义了消息传输过程中要用到的数据结构。消息资源编译成功后，也会像图片资源和铃声资源一样，有一个文件列出系统中所有的消息资源以及消息资源的处理函数，可在 plutommi\customer\customerinc\mmi_rp_callback_mgr_config.h 文件中查看。

接下来，我们使用消息实现一个小功能，当主菜单中高亮显示 MAIN_MENU_MYSTUDY_ID 菜单时，就在屏幕上打印出"I am highlight message"。

在 MyStudyAppMain.res 文件中添加消息资源，代码只有两行，如代码清单 3.1-12 所示。添加之后，就可以使用 make resgen 编译资源了。虽然 EVT_ID_SRV_MYSTUDY_MSG_IND 消息 id 和 mmi_mystudy_msg_proc 函数都还没有定义，但编译资源的时候不会报错，只是编译模拟器会报错。编译之后，可以到 mmi_rp_callback_mgr_config.h 文件中查看是否编译成功，读者也可以看看 mmi_rp_app_mystudy_def.h 文件中是否有生成消息资源 id 对应的枚举类型。

--代码清单 3.1-12--
```
#if defined(__MYSTUDY_APPLICATION__)
#include "custresdef.h"

/* Need this line to tell parser that XML start, must after all #include. */
```

```xml
<?xml version="1.0" encoding="UTF-8"?>

<APP id="APP_MYSTUDY">
    /*省略部分代码*/
    <!----------------audio resource---------------------------------------------------------->
    <AUDIO id="AUD_MY_MUSIC_ID" flag="MULTIBIN">CUST_ADO_PATH
            "\\\\MyStudyApp\\\\music.mp3"</AUDIO>


    <!----------------NV 资源---------------------------------------------------------->
    <CACHEDATA type="byte" id="NVRAM_MYSTUDY_ID" restore_flag="TRUE">
        <DEFAULT_VALUE> [0X00] </DEFAULT_VALUE>
    </CACHEDATA>


    <!----------------定时器资源-------------------------------------->
    <TIMER id="MYSTUDY_TIMER_ID"/>


    <!----------------消息资源-------------------------------------->
    <SENDER id="EVT_ID_SRV_MYSTUDY_MSG_IND" hfile="MyStudyAppMain.h"/>
    <RECEIVER id="EVT_ID_SRV_MYSTUDY_MSG_IND" proc="mmi_mystudy_msg_proc"/>


    <!----------------屏幕资源-------------------------------------->
    <SCREEN id="SCR_ID_MYSTUDY_MENU_LIST"/>
    <SCREEN id="SCR_ID_HELLO_WORLD"/>

</APP>
#endif
```

在 MyStudyAppMain.h 文件中定义消息资源 id、消息类型、消息的数据结构,代码如清单 3.1-13 所示。其中 MYSTUDY_BASE 是在 mmi_pluto_res_range_def.h 文件中定义的,注意需包含头文件 MMIDataType.h。

---代码清单 3.1-13---

```c
#ifndef __MYSTUDYAPPMAIN_H__
#define __MYSTUDYAPPMAIN_H__
#if defined(__MYSTUDY_APPLICATION__)
#include "MMIDataType.h"

/*消息资源 id*/
typedef enum
{
```

```
    EVT_ID_SRV_MYSTUDY_MSG_IND = MYSTUDY_BASE,
    EVT_ID_SRV_MYSTUDY_MSG_MAX
}srv_mystudy_msg_id;

/*消息类型*/
typedef enum
{
    SRV_MYSTUDY_MSG_NONE = 0,
    SRV_MYSTUDY_MSG_HIGHLIGHT,
    SRV_MYSTUDY_MSG_MAX
}srv_mystudy_msg_type;

/*消息数据*/
typedef struct
{
    MMI_EVT_PARAM_HEADER
    U8 type;
}stu_mystudy_msg_data;

extern void mtk_helloworld(void);
#endif
#endif
```

在 MyStudyAppMain.c 文件中定义消息处理函数 mmi_mystudy_msg_proc，并在 MAIN_MENU_MYSTUDY_ID 菜单的高亮函数 mmi_highlight_mystudy 中发送消息。发送消息分为三步：第一步调用 MMI_FRM_INIT_EVENT 宏函数初始化消息数据；第二步设置消息数据；第三步调用 MMI_FRM_CB_EMIT_POST_EVENT 函数发送消息。这三步是固定的，我们只需记住如何使用。代码如代码清单 3.1-14 所示，注意需包含头文件 MyStudyAppMain.h 和 mmi_cb_mgr_gprot.h。

---------------------------------------代码清单 3.1-14---------------------------------------

```
/*省略部分代码*/
void mmi_highlight_mystudy(void)
{
    stu_mystudy_msg_data evt={0x00};

    /*发送消息*/
    MMI_FRM_INIT_EVENT(&evt, EVT_ID_SRV_MYSTUDY_MSG_IND);
    evt.type = SRV_MYSTUDY_MSG_HIGHLIGHT;/*消息类型 id*/
    MMI_FRM_CB_EMIT_POST_EVENT((mmi_event_struct *)&evt);
```

```
        SetKeyHandler(mmi_entry_mystudy, KEY_LSK, KEY_EVENT_UP);
}

/*消息处理函数*/
mmi_ret mmi_mystudy_msg_proc(mmi_event_struct *evt)
{
        stu_mystudy_msg_data *event = (stu_mystudy_msg_data*)evt;

        if(EVT_ID_SRV_MYSTUDY_MSG_IND == evt->evt_id)
        {
                if(SRV_MYSTUDY_MSG_HIGHLIGHT==event->type)/*处理 highlight 消息*/
                {
                        gui_set_text_color(UI_COLOR_RED);
                        gui_move_text_cursor(10, 100);
                        gui_set_font(&MMI_large_font);
                        gui_print_text(L"I am highlight message");
                        gui_BLT_double_buffer(0, 0, UI_DEVICE_WIDTH, UI_DEVICE_HEIGHT);
                }
        }
        return MMI_RET_OK;
}
```

运行模拟器，当高亮显示最后一个主菜单时，就可以看到如图 3.1-18 所示的效果，读者可在 mmi_highlight_mystudy 函数和 mmi_mystudy_msg_proc 函数打上断点，跟踪代码的具体执行流程。

以上代码虽然在模拟器中编译通过了，但是在 make r mmiresource 的时候，会提示 mmi_rp_callback_mgr_header_file.h 文件中找不到 MyStudyAppMain.h 头文件，这是因为 mmiresource 模块中没有导入 MyStudyApp 模块的头文件路径。参考 mmi_app 模块的 Makefile 文件 mmi_app.mak 中导入 MyStudyApp 头文件的方法，把代码复制到 mmiresource.mak 文件的最后面。

图 3.1-18

3.2 绘制界面

屏幕上显示的任何元素，都是通过代码绘制出来的。在 MTK 中有一套完整的界面绘制函数，这些函数大部分都以 gui 开头，比如前面例子中用到的 gui_print_text 函数，此外还有一些以 gdi 开头的界面绘制函数。gui 函数与 gdi 函数的区别在于 gdi 函数会使用硬件加速。但我们在开发中不必纠结该使用 gdi 函数还是 gui 函数,因为系统已经为我们分配好了,必须使用 gdi 函数的地方是没有 gui 接口的。MTK 界面绘制的接口主要包括字符串、图片的显示以及字符串和图片尺寸的测量等一些功能性函数,另外还有绘制直线、矩形、圆形以及填充矩形区域的一些函数。灵活使用这些函数接口,配合图片、动画可以绘制出相当生动的画面。界面绘制的函数接口都处在 framework 层。

3.2.1 清屏

在前面的实例中,我们虽然把字符"hello world!"显示在屏幕上,但同时屏幕上还有其他的内容,而这些内容并不是我们想要的。前面已经说过,MTK 屏幕上显示的任何内容都是我们自己绘制的,所以我们只能通过内容覆盖的方式清除界面上不需要的显示内容。

修改 mtk_helloworld 函数,添加一行代码,在显示字符串之前,把整个屏幕填充成黑色作为背景色,用来覆盖掉上一个界面的显示内容,并把字符显示的颜色改为白色。代码如下(代码清单 3.2-1):

```
--------------------------------------------代码清单 3.2-1--------------------------------------------
/*省略部分代码*/
void mtk_helloworld(void)
{
    mmi_frm_scrn_enter(GRP_ID_ROOT, SCR_ID_HELLO_WORLD, mtk_helloworld_exit,
                mtk_helloworld, MMI_FRM_FULL_SCRN);
    gui_fill_rectangle(0, 0, UI_device_width, UI_device_height, UI_COLOR_BLACK)
                /*设置背景色为黑色*/
    gui_set_text_color(UI_COLOR_WHITE);      /*设置字符颜色为白色*/
    gui_move_text_cursor(10, 10);            /*移动字符显示坐标为*/
    gui_set_font(&MMI_large_font);           /*设置字符显示的字体*/
    gui_print_text((UI_string_type)GetString(STR_ID_HELLO_WORLD));/*在屏幕上打印字符*/
    kal_prompt_trace(MOD_XDM, "-mtk_helloworld-%d--%s--", __LINE__, __FILE__);
    gui_BLT_double_buffer(0, 0, UI_DEVICE_WIDTH, UI_DEVICE_HEIGHT);
}
/*省略部分代码*/
--------------------------------------------------------------------------------------------------
```

gui_fill_rectangle 函数的功能是填充矩形区域,前面四个参数分别为矩形左上角和右下角的 x、y 坐标,最后一个参数是填充矩形的颜色,其中 UI_device_width 和 UI_device_height 是两个全局变量,分别表示屏幕的宽度和高度,另外还有两组宏定义 UI_DEVICE_WIDTH、UI_DEVICE_HEIGHT 和 LCD_WIDTH、LCD_HEIGHT 也是表示屏幕的宽度和高度,可以根据自己的喜好选择使用。在绘制界面的时候，需要特别注意 MTK 的屏幕坐标系原点位于左上角，最大区域点位于右下角。把 mmi_highlight_mystudy 函数中发送消息的代码注释掉，只保留最后一行代码。运行模拟器，此时显示的界面如图 3.2-1 所示。

现在屏幕上显示的所有内容都是我们自己设置的。我们使用填充矩形区域的方式，给屏幕设置了一个背景色，这种设置背景色的操作，我们称之为"清屏"，就是清除屏幕上的所有内容，只保留单一的一种颜色。清屏的函数有多个，比较常用的除了 gui_fill_rectangle 外，还有 GDI_RESULT

图 3.2-1

gdi_layer_clear(gdi_color bg_color)，这个函数是层相关的接口，它只有一个背景颜色的参数，把当前层的区域刷成背景色。另外还有一个填充矩形区域的函数接口 void gui_draw_filled_area(S32 x1, S32 y1, S32 x2, S32 y2, UI_filled_area *f)，这个函数接口普遍用于主题相关的区域填充，比如顶部的状态栏、标题栏，底部的软键栏，还有系统的一些背景色都是使用这个接口进行填充的。读者可以分别使用以下两条语句替换 gui_fill_rectangle 语句，看看运行效果。

(1) gdi_layer_clear(GDI_COLOR_BLACK);　　(注意包含头文件的 "gdi_include.h")

(2) gui_draw_filled_area(0, 0, UI_device_width, UI_device_height,
　　　　　　　　current_MMI_theme->sub_menu_bkg_filler);

第(2)条语句的运行效果，背景颜色取决于当前主题设置的颜色，我们可以通过修改主题的方式来改变背景色，这方面的内容我们在后面介绍主题的时候再讲解。

3.2.2 显示图片

图片也是界面显示的主要内容，在讲解资源的章节中我们提到过，MTK 中显示的资源有三种，即图片资源、图片文件和图片二进制数据，其中图片资源是用得最多的，图片文件主要用于相册的浏览，而直接显示图片二进制数据的方法目前很少使用。本节主要介绍图片资源的显示方法。

继续修改 mtk_helloworld 函数，在倒数第二行添加一行显示图片的代码 gdi_image_draw(10, 40, GetImage(IMG_STATIC_ID));具体如下所示(代码清单 3.2-2)：

---代码清单 3.2-2---

```
/*省略部分代码*/
void mtk_helloworld(void)
{
    mmi_frm_scrn_enter(GRP_ID_ROOT, SCR_ID_HELLO_WORLD, mtk_helloworld_exit,
              mtk_helloworld, MMI_FRM_FULL_SCRN);
    gui_fill_rectangle(0, 0, UI_device_width, UI_device_height, UI_COLOR_BLACK);
                /*设置背景色为黑色*/
```

```
        gui_set_text_color(UI_COLOR_WHITE);          /*设置字符颜色为白色*/
        gui_move_text_cursor(10, 10);                 /*移动字符显示坐标为*/
        gui_set_font(&MMI_large_font);                /*设置字符显示的字体*/
        gui_print_text((UI_string_type)GetString(STR_ID_HELLO_WORLD));/*在屏幕上打印字符*/

        kal_prompt_trace(MOD_XDM, "-mtk_helloworld-%d--%s--", __LINE__, __FILE__);

        gdi_image_draw(10, 40, GetImage(IMG_STATIC_ID)); /*显示静态图片*/

        gui_BLT_double_buffer(0, 0, UI_DEVICE_WIDTH, UI_DEVICE_HEIGHT);
    }
    /*省略部分代码*/
```

这行代码是在屏幕上坐标(10，40)位置处显示我们之前新增的静态图片资源 IMG_STATIC_ID，运行模拟器，效果如图 3.2-2 所示。

MTK 中关于图片的函数接口，命名规则中大多包含 "image" 字符，这些函数接口大部分都在 gdi_include.h 头文件中可以找到其原型。通过函数名称我们就能大致猜出这个函数是用来干什么的，至于如何使用它，前面已经介绍过，我们可以从 MTK 的原始代码中学习。

上述代码中 gdi_image_draw 函数就是用来显示图片的，前面两个参数是图片显示的左上角坐标，最后一个参数是图片的

图 3.2-2

数据指针，通过 S8 *GetImage(U16 ImageId)函数可以获取图片资源 id 对应的二进制数据指针。关于图片资源显示的函数还有另外两个，读者可以分别用以下两行代码代替 gdi_image_draw 来显示图片，看看运行效果。

```
        gdi_image_draw_id(10, 40, IMG_STATIC_ID);
        gdi_image_draw_resized_id(10, 40, 200, 200, IMG_STATIC_ID);
```

第一行代码显示的效果并没有任何变化，这个函数只需要传入图片资源 id，实际上就相当于封装了 gdi_image_draw 函数和 GetImage 函数。而第二行代码显示的图片比之前的要大些，这个函数可以按指定尺寸显示图片，实现图片的缩放效果，图片的原始尺寸为 60×60，放大后显示的尺寸为 200×200。

以上介绍的是静态图片资源的显示方法，在资源章节中，我们还添加了一个动画资源。在 MyStudyAppMain.c 文件中修改代码如下(代码清单 3.2-3)：

--代码清单 3.2-3---
```
    /*省略部分代码*/
    static gdi_handle g_anim_handle = NULL;/*动画句柄*/
    void mtk_helloworld_exit(void)
    {
```

```
        gdi_anim_stop(g_anim_handle);/*停止动画*/
}
void mtk_helloworld(void)
{
        mmi_frm_scrn_enter(GRP_ID_ROOT, SCR_ID_HELLO_WORLD, mtk_helloworld_exit,
                mtk_helloworld, MMI_FRM_FULL_SCRN);
        gui_fill_rectangle(0, 0, UI_device_width, UI_device_height, UI_COLOR_BLACK);
                /*设置背景色为黑色*/

        gui_set_text_color(UI_COLOR_WHITE);        /*设置字符颜色为白色*/
        gui_move_text_cursor(10, 10);              /*移动字符显示坐标为*/
        gui_set_font(&MMI_large_font);             /*设置字符显示的字体*/
        gui_print_text((UI_string_type)GetString(STR_ID_HELLO_WORLD));/*在屏幕上打印字符*/

        kal_prompt_trace(MOD_XDM, "-mtk_helloworld-%d--%s--", __LINE__, __FILE__);

         gdi_image_draw(10, 40, GetImage(IMG_STATIC_ID));         /*显示静态图片*/
        gdi_anim_draw_id(10,  100,  IMG_ANIM_ID, &g_anim_handle);  /*显示动画*/

        gui_BLT_double_buffer(0, 0, UI_DEVICE_WIDTH, UI_DEVICE_HEIGHT);
}
/*省略部分代码*/
```

动画的显示稍微复杂点，需要先定义一个全局变量作为动画句柄，然后调用 gdi_anim_draw_id 函数显示动画，最后在退出屏幕时，我们需要调用 gdi_anim_stop 函数使动画停止，否则它会一直显示在界面中，除非系统调用了 gdi_anim_stop_all 函数停止了所有动画。以上代码的运行效果如图 3.2-3 所示。

关于动画显示的一些函数接口也可以在 gdi_include.h 文件中找到，它们的命名规则中通常包含"anim"字符。动画显示与图片类似，可以使用显示图片的接口显示动画，但只会显示动画的第一帧，也可以使用动画的显示接口显示图片。另外，在动画播放过程中可以设置一些回调函数处理其他事情，比如函数 GDI_RESULT gdi_anim_set_draw_before_callback(void (*callback_ptr)(GDI_RESULT result)) 设置动画播放之前的回调函数，函数 GDI_RESULT gdi_anim_set_last_frame_callback(void (*callback_ptr) (GDI_RESULT result))设置播放完最后一帧的回调函数，函数 GDI_RESULT gdi_anim_set_draw_after_callback(void (*callback_ptr) (GDI_RESULT result))设置播放完成后的回调函数。这些函数的使用方法，在后面的游戏示例代码中会使用到。

图 3.2-3

3.2.3 界面排版

上面显示的字符、图片和动画,我们都是随便指定的坐标,这样显示的效果很不美观。在实际开发中,为了界面的美观,我们经常需要按特定效果显示,比如居中、靠左、靠右等。接下来我们把字符在 x 坐标上居中,在 y 坐标上靠上显示(y=5),作为这个页面的标题,并把动画和图片显示在同一行,相隔 10 个像素,然后水平居中,y 轴上显示坐标在字符下面 20 个像素。修改 mtk_helloworld 函数,代码如清单 3.2-4 所示。

```
---------------------------------------代码清单 3.2-4---------------------------------------
void mtk_helloworld(void)
{
    S32 str_x=0, str_y=0, str_w=0, str_h=0;
    S32 img_x=0, img_y=0, img_w=0, img_h=0, anim_w=0, anim_h=0;

    mmi_frm_scrn_enter(GRP_ID_ROOT, SCR_ID_HELLO_WORLD, mtk_helloworld_exit,
                    mtk_helloworld, MMI_FRM_FULL_SCRN);
    gui_fill_rectangle(0, 0, UI_device_width, UI_device_height, UI_COLOR_BLACK);
                        /*设置背景色为黑色*/
    //gdi_layer_clear(GDI_COLOR_BLACK);
    //gui_draw_filled_area(0,      0,       UI_device_width,        UI_device_height,
      current_MMI_theme->sub_menu_bkg_filler);

    gui_set_text_color(UI_COLOR_WHITE);/*设置字符颜色为白色*/
    gui_measure_string((UI_string_type)GetString(STR_ID_HELLO_WORLD), &str_w, &str_h);
                        /*测量字符宽度和高度*/
    str_x = (UI_DEVICE_WIDTH-str_w)/2;
    str_y = 5;
    gui_move_text_cursor(str_x, str_y);     /*移动字符显示坐标*/
    gui_set_font(&MMI_large_font);          /*设置字符显示的字体*/
    gui_print_text((UI_string_type)GetString(STR_ID_HELLO_WORLD));/*在屏幕上打印字符*/
    kal_prompt_trace(MOD_XDM, "-mtk_helloworld-%d--%s--", __LINE__, __FILE__);
    gdi_image_get_dimension_id(IMG_STATIC_ID, &img_w, &img_h);
    gdi_image_get_dimension_id(IMG_ANIM_ID, &anim_w, &anim_h);
    img_x = (LCD_WIDTH - (img_w+anim_w+10))/2;
    img_y = str_y+str_h+20;
    gdi_image_draw(img_x, img_y, GetImage(IMG_STATIC_ID));       /*显示图片*/
//  gdi_image_draw_id(10, 40, IMG_STATIC_ID);
//  gdi_image_draw_resized_id(10, 40, 200, 200, IMG_STATIC_ID);    /*图片缩放显示*/
    gdi_anim_draw_id(img_x+img_w+10, img_y, IMG_ANIM_ID, &g_anim_handle);/*显示动画*/
    gui_BLT_double_buffer(0, 0, UI_DEVICE_WIDTH, UI_DEVICE_HEIGHT);
```

}

界面排版实际上就是显示坐标的计算,在计算过程中需要知道屏幕的尺寸,前面已经介绍过,屏幕的尺寸有三种获取方式(UI_device_width, UI_device_height)、(UI_DEVICE_WIDTH, UI_DEVICE_HEIGHT)、(LCD_WIDTH,LCD_HEIGHT),可根据自己的喜好随便选择。另外还需要知道显示内容的尺寸,在以上代码中使用了两个测量尺寸的函数:gui_measure_string 的函数功能是用于测量字符串的宽度和高度,其函数原型可在 gui.h 文件中查看;gdi_image_get_dimension_id 函数用于测量图片的宽度和高度,其函数原型可在 Gdi_include.h 文件中查看。另外还有一个函数 gui_measure_image 也可以测量图片尺寸,功能是一样的,只不过要以图片数据指针作为函数参数。

有了内容尺寸和屏幕尺寸,我们就可以准确地计算出显示坐标了。运行以上代码,效果如图 3.2-4 所示。

图 3.2-4

3.2.4 绘制几何图形

在前面讲解清屏的时候,我们利用填充矩形区域的方法设置背景色。实际上 MTK 屏幕上的背景色都是使用这种方式设置的。除了填充矩形区域的接口外,MTK 系统还提供了其他的一些几何图形绘制及填充函数,这些函数都包含在 gdi_include.h 文件中,而有的函数以函数指针的形式存在其他的形态,包含在别的头文件中,比如 gui_fill_rectangle,它是一个函数指针,指向 UI_fill_rectangle 函数,而 UI_fill_rectangle 函数中调用的依旧是 gdi_draw_solid_rect 函数。

继续修改 mtk_helloworld 函数,我们将界面的标题背景填充成绿色,然后在界面上绘制一个正方形和一个圆形,代码如下(代码清单 3.2-5):

---------------------------------------代码清单 3.2-5---------------------------------------
```
/*省略部分代码*/
void mtk_helloworld(void)
{
    S32 str_x=0, str_y=0, str_w=0, str_h=0;
    S32 img_x=0, img_y=0, img_w=0, img_h=0, anim_w=0, anim_h=0;

    mmi_frm_scrn_enter(GRP_ID_ROOT, SCR_ID_HELLO_WORLD, mtk_helloworld_exit,
                    mtk_helloworld, MMI_FRM_FULL_SCRN);
    gui_fill_rectangle(0, 0, UI_device_width, UI_device_height, UI_COLOR_BLACK);
                    /*设置背景色为黑色*/
    //gdi_layer_clear(GDI_COLOR_BLACK);
    //gui_draw_filled_area(0, 0, UI_device_width, UI_device_height,
                    current_MMI_theme->sub_menu_bkg_filler);
```

```
gui_set_text_color(UI_COLOR_WHITE);/*设置字符颜色为白色*/
gui_measure_string((UI_string_type)GetString(STR_ID_HELLO_WORLD), &str_w, &str_h);
                  /*测量字符宽度和高度*/
str_x = (UI_DEVICE_WIDTH-str_w)/2;
str_y = 5;
gui_move_text_cursor(str_x, str_y);        /*移动字符显示坐标为*/
gui_set_font(&MMI_large_font);             /*设置字符显示的字体*/
gdi_draw_solid_rect(0, 0, UI_DEVICE_WIDTH, str_y+str_h+5, GDI_COLOR_GREEN);
gui_print_text((UI_string_type)GetString(STR_ID_HELLO_WORLD));/*在屏幕上打印字符*/
kal_prompt_trace(MOD_XDM, "-mtk-helloworld-%d--%s--", __LINE__, __FILE__);
gdi_image_get_dimension_id(IMG_STATIC_ID, &img_w, &img_h);
gdi_image_get_dimension_id(IMG_ANIM_ID, &anim_w, &anim_h);
img_x = (LCD_WIDTH - (img_w+anim_w+10))/2;
img_y = str_y+str_h+20;
gdi_image_draw(img_x, img_y, GetImage(IMG_STATIC_ID));        /*显示图片*/
//   gdi_image_draw_id(10, 40, IMG_STATIC_ID);
//   gdi_image_draw_resized_id(10, 40, 200, 200, IMG_STATIC_ID);   /*图片缩放显示*/
gdi_anim_draw_id(img_x+img_w+10, img_y, IMG_ANIM_ID, &g_anim_handle);/*显示动画*/
gdi_draw_circle(60, 160, 30, GDI_COLOR_RED);
gdi_draw_rect(150, 130, 210, 180, GDI_COLOR_RED);
gui_BLT_double_buffer(0, 0, UI_DEVICE_WIDTH, UI_DEVICE_HEIGHT);
}
```

运行模拟器，显示效果如图 3.2-5 所示。

图 3.2-5

几何图形除了矩形和圆形外，还有直线、椭圆、多边形等，所有几何图形的实现都依赖于一个描点函数：void gdi_draw_point(S32 x, S32 y, gdi_color pixel_color)。只要知道绘制几何图形的数学公式，我们就完全可以定制任何图形的绘制接口。

3.2.5 给界面添加背景音乐

界面中有个阿狸在跳舞，我们来给它加个背景音乐。在讲解资源的章节中，已经添加了一个铃声资源 AUD_MY_MUSIC_ID，我们就用它来做界面的背景音乐。音频相关的函数接口都在 plutommi\Service\MDI\MDISrc\mdi_audio.c 文件中。播放铃声资源要用到以下三个函数：

(1) U8 *get_audio(MMI_ID_TYPE i, U8 *type, U32 *filelen)。

这个函数用于获取铃声资源的数据指针。第一个参数(i)就是我们在资源中定义的铃声资源 id；第二个参数是铃声资源文件的格式，通常铃声文件格式有 MP3、wav 等；第三个参数是铃声资源文件的长度。

(2) mdi_result mdi_audio_play_string_with_vol_path(void *audio_data, U32 len, U8 format, U8 play_style, mdi_ext_callback handler, void *user_data, U8 volume, U8 path)。

这个函数用于播放铃声资源。第一个参数(audio_data)是音频资源的数据指针，即 get_audio 函数的返回值；第二个参数(len)表示铃声资源文件的长度，即 get_audio 函数第三个参数 filelen 中获取的值；第三个参数(format)表示音频资源文件的格式，即 get_audio 函数第二个参数 type 中获取的值；第四个参数(play_style)表示播放的类型，取值见 interface\ps\include\device.h 文件中的枚举 audio_play_style_enum，比如无限循环播放(DEVICE_AUDIO_PLAY_INFINITE)、单次播放(DEVICE_AUDIO_PLAY_ONCE)；第五个参数(handler)是播放音频的回调函数，如果不需要可以传入 NULL；第六个参数(user_data)是用户数据，同样可以传入 NULL；第七个参数(volume)是铃声播放的声音大小，根据音量等级不同，需要用宏函数转换，比如使用 7 级音量，取值可见 plutommi\Service\MDI\MDIInc\mdi_audio.h 文件的枚举 MDI_AUD_VOL_ENUM，则可传入参数 MDI_AUD_VOL_MUTE(MDI_AUD_VOL_6)；第八个参数(path)是声音播放的路径，比如从耳机播放(MDI_DEVICE_SPEAKER2)或从喇叭播放(MDI_DEVICE_LOUDSPEAKER)，又或者同时播放(MDI_DEVICE_SPEAKER_BOTH)，这个参数只在真机上才起作用。

(3) mdi_result mdi_audio_stop_all(void)。

这个函数的功能是停止所有声音的播放。因为播放铃声资源时，我们通常不能直接操作铃声资源 id，所有可以调用这个函数来停止自己播放的铃声。

在 MyStudyAppMain.c 文件中，新增一个函数 mmi_mystudy_audio_play 用于播放铃声资源，并放到 mtk_helloworld 函数中调用，当退出屏幕时在 mtk_helloworld_exit 函数中停止铃声，修改代码如下（代码清单 3.2-6）：

```
-------------------------------------代码清单 3.2-6-------------------------------------
/*省略部分代码*/
void mmi_mystudy_audio_play(void)
{
    /*----------------------------------------------------------------*/
    /* Local Variables                                                */
    /*----------------------------------------------------------------*/
```

```c
    mdi_result play_result = MDI_AUDIO_FAIL;
    U32 audio_len = 0;
    U8 *audio_data=NULL, audio_type=0;

    /*----------------------------------------------------------------*/
    /* Code Body                                                      */
    /*----------------------------------------------------------------*/
    audio_data = get_audio(AUD_MY_MUSIC_ID, &audio_type, &audio_len);
    play_result = mdi_audio_play_string_with_vol_path(
              (void*)audio_data,
              audio_len,
              audio_type,
              DEVICE_AUDIO_PLAY_ONCE,
              NULL,
              NULL,
              MDI_AUD_VOL_MUTE(MDI_AUD_VOL_6),
              MDI_AUD_PTH_EX(MDI_DEVICE_SPEAKER_BOTH));
}

void mtk_helloworld_exit(void)
{
    gdi_anim_stop(g_anim_handle);/*停止动画*/
    g_anim_handle = NULL;
    mdi_audio_stop_all();
}
void mtk_helloworld(void)
{
    /*省略部分代码*/
    mmi_mystudy_audio_play();

    gui_BLT_double_buffer(0, 0, UI_DEVICE_WIDTH, UI_DEVICE_HEIGHT);
}
/*省略部分代码*/
```

运行模拟器，当进入 SCR_ID_HELLO_WORLD 屏幕时，音乐就会响起，退出时则会停止。铃声资源在物联网或智能手表的开发过程中用得不多，主要是定制系统的铃声资源，比如开关机铃声、来电铃声、闹钟铃声等。

3.2.6 定制屏幕尺寸

MTK 系统中已经兼容了主流的屏幕尺寸，我们可以在 make 目录下的项目配置文件 HEXING60A_11B_GPRS.mak 中设置 MAIN_LCD_SIZE 宏的属性值来配置 MTK 设备的屏幕尺寸，本书所示的代码使用的屏幕尺寸为 240×320，代码如下：

```
MAIN_LCD_SIZE=240X320
        # Add a MAIN_LCD_SIZE compile option in make file
        # 320X240：The main lcd size is 320X240
        # 240X320：The main lcd size is 240X320
        # 128X128：The main lcd size is 128X128
        # 96X64：The main lcd size is 96X64
        # 360X640：The main lcd size is 360X640
        # 128X160：The main lcd size is 128X160
        # 176X220：The main lcd size is 176X220
        # 240X400：The main lcd size is 240X400
        # 320X480：The main lcd size is 320X480
        # Switch-ability：
        # 128X128： -> 128X160
```

在 MAIN_LCD_SIZE 宏定义的下方，使用"#"注释的内容中，还有其他的一些屏幕尺寸，比如 320×240、128×128 等，这些屏幕尺寸都是 MTK 系统默认且已经兼容的，我们只需拿来用就可以了。系统中已经兼容的这些屏幕尺寸都是手机上的主流屏幕尺寸，如果我们开发一款智能手表，它的屏幕尺寸不可能使用 240×320 这么大。现在市场上主流的智能手表屏幕尺寸有 240×240、128×128、128×64、96×64、64×48 等，都是一些尺寸较小的屏幕，这些屏幕尺寸很多都是 MTK 系统默认不支持的，需要我们自己定制。

其实自己定制屏幕尺寸也挺简单的，我们只需要把 MAIN_LCD_SIZE 设置为指定的屏幕尺寸，比如 240×240，然后使用 make new 指令编译代码，在编译的过程中肯定会报错，接下来的过程就是不断地修改编译错误。因为系统中的很多功能都与屏幕尺寸相关，不同的屏幕尺寸，界面显示也有差异。每个屏幕尺寸都会生成一个对应的宏，比如 _MMI_MAINLCD_240X240_、_MMI_MAINLCD_240X320_，我们用这些宏控制代码实现条件编译，从而可以实现多个屏幕尺寸的兼容。

在本书的 MTK6260A 系统中，笔者已经定制了 240X240 屏幕尺寸。读者可在 Source Insight 中全局查找 _MMI_MAINLCD_240X240_ 和 240X240 来查看代码修改的地方。除了修改代码之外，还有一些与屏幕尺寸相关的文件和目录需要创建。创建的方法可以拷贝一份相近屏幕尺寸的内容，然后重命名，再修改里面的文件内容以适配新增的屏幕尺寸。比如在 plutommi\Customer\Images 目录复制一份 PLUTO240X320 的图片资源，然后重命名为 PLUTO240X240，就成了 240X240 屏幕尺寸的图片资源目录了。还有主题文件 plutommi\Customer\LcdResource\MainLcd240X240 目录、模拟器皮肤 MoDIS_VC9\MoDIS\Skin\Normal\240X240 目录，也可以使用同样的方法创建。

代码 make new 编译通过之后,再使用 make gen_modis 重新生成模拟器,运行模拟器如图 3.2-7 所示。

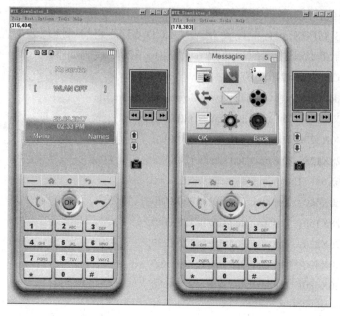

图 3.2-7

在主菜单界面里面已经找不到我们自己添加的菜单了,因为我们的主菜单添加在 MainMenuRes.res 文件中的_MMI_MAINLCD_240X320_代码分支下。如果要新增到 240×240 屏幕主菜单中,需在默认的主菜单分支下也新增一份,这个功能留给读者自己完成。在后续的示例代码中,我们依旧会使用 240×320 的屏幕尺寸配置。

3.3 按　　键

在前面的示例程序中,我们已经多次接触过按键了。按键的使用其实挺简单的,我们在 MMI 层只需要注册它执行的功能就可以了。

每一个按键在不同的屏幕中都有不同的功能,这些功能都可以自己定义。如果进入一个屏幕,没有指定按键功能,则这个按键默认是没有功能的,比如我们上一章节绘制的界面中,按右软键"KEY_RSK"是没有任何反应的。按键注册的有效区域只限于当前屏幕,如果退出当前屏幕,则按键功能会被清空,进入下一个屏幕时又会被注册成其他的功能。这样我们在不同的界面,按相同的按键所看到的效果是不一样的。

模拟器上涵盖了 MTK 上所有常用的按键,我们把所有的按键分为功能按键和数字按键。功能按键类似于左软键(KEY_LSK)、右软键(KEY_RSK)、开关机键(KEY_END)还有音量增减按键和上下左右方向按键等,这些按键在不同的界面通常都可以注册不同的功能。数字按键主要是 0~9,这些按键主要用于拨号使用,但在 MTK 物联网设备中这些按键基本上已经被淘汰了。另外还有*、#按键既可以做功能按键也可以做数字按键。MMI 层使用的按键键值都定义在 GlobalConstants.h 文件的 mmi_keypads_enum 枚举中,并在 KeyBrd.c

文件的 g_key_code_map 表中实现与物理按键的映射关系。物理按键的定义需要根据硬件原理图来设计，不同的设备差异很大，本书不作讲解。

每个按键都有对应的按键事件，常用的按键事件有三种：按下(KEY_EVENT_DOWN)、弹起(KEY_EVENT_UP)、长按(KEY_LONG_PRESS)。按下和弹起事件很好理解，如果让按键处于按下的状态持续一段时间(约 3 秒)就会触发长按事件。这些事件 id 定义在 GlobalConstants.h 文件的 mmi_key_types_enum 枚举中。还有另外一些按键事件，我们基本上用不到，有一些 MTK 平台上也没有实现，只是留了一个接口而已。我们在注册按键功能的时候，还要指定按键的事件，同一个按键在不同的按键事件下可以执行不同的按键功能。

注册按键功能，通常都是使用 SetKeyHandler 函数，这个函数实际上是对 mmi_frm_set_key_handler 函数进行了一层封装，它对于任一按键的注册都是通用的。函数说明如下：

    void SetKeyHandler (funcPtr funcPtr, U16 keyCode, U16 keyType)

其中，funcPtr：按键执行的功能函数指针；

  keyCode：按键值，取值范围见 GlobalConstants.h 文件中的枚举 mmi_keypads_enum；

  keyType：按键方式，取值范围见 GlobalConstants.h 文件中的枚举 mmi_key_types_enum。

还有一些函数只是针对特定按键进行注册，比如左软键、右软键、中间键。我们在调用这些函数的时候，只需要传入按键执行的函数及对应的事件即可，其实现方式相当于对 SetKeyHandler 函数再进行了一层封装，只不过指定了 keyCode 的值。函数原型如下：

（1）注册左软键：void SetLeftSoftkeyFunction(void (*f) (void), MMI_key_event_type k)。

（2）注册右软键：void SetRightSoftkeyFunction(void (*f) (void), MMI_key_event_type k)。

（3）注册中间键：void SetCenterSoftkeyFunction(void (*f) (void), MMI_key_event_type k)。

以上的函数接口都是用于注册单个按键，还有一个接口同时注册多个按键，当我们按下其中任一按键都可以执行对应的函数接口。函数说明如下：

    void SetGroupKeyHandler(funcPtr funcPtr, PU16 keyCodes, U8 len, U16 keyType)

其中，funcPtr：按键执行的功能函数指针；

  keyCodes：按键值数组，取值范围见 GlobalConstants.h 文件中的枚举 mmi_keypads_ enum；

  keyType：按键方式，取值范围见 GlobalConstants.h 文件中的枚举 mmi_key_types_enum。

在 idle 界面按 0～9 任一数字按键时，都会进入拨号界面，就是通过这个函数实现的。但在系统其他的地方都没有再使用到这个函数。我们在实际开发中也基本上用不到，特别是 MTK 物联网设备，有的没有按键，有的只有 1 个按键，用户操作比较频繁的智能手表，目前也没发现超过 5 个按键的。

按键虽然是一个比较重要的内容，但是它却非常简单，在后面的代码中，我们会多次使用它。现在我们在 mtk_helloworld 函数的最后面添加如下一行代码，注册右软键的弹起事件执行退出当前屏幕，当按下右软键时会返回上一级屏幕，也就是我们自定义的菜单列表界面。请读者自己运行看看效果。

    SetRightSoftkeyFunction(mmi_frm_scrn_close_active_id, KEY_EVENT_UP);

第四章　MMI 高级编程

4.1　定　时　器

定时器的作用是规定程序多长时间之后或者每隔多长时间做什么动作。定时器在 MTK 中应用比较广泛，基于定时器开发的功能有闹钟、秒表、定时开关机等，动画的播放实际上就是通过定时器实现的。我们在后面讲解网络的时候，也要用到定时器。定时器的类型我们归为两种：一种是自定义的定时器，可以自己控制停止和运行，计时时间可以精确到毫秒，我们就称之为定时器；而另一种是系统定时器，每个 MTK 设备上都有 RTC(Real-Time Clock 实时时钟)，只要设备上电就会启动运行，并且无法在代码中把它停止，系统中所有时间相关的功能都是基于 RTC 实现的。我们也可以把它当做一个定时器来使用，MTK 系统中把这个定时器称之为 reminder，闹钟和定时开关机的功能就是基于此实现的。因为它每次更新时间的时候，都会发送中断消息，所以我们也可以把这个定时器称之为"RTC 中断"，我们能检测到的 RTC 中断间隔为一分钟。

4.1.1　普通定时器

普通定时器的定义方式有两种：一种是通过资源加载；另一种是定义在 TimerEvents.h 文件的 MMI_TIMER_IDS 枚举中。在编写 MMI 层应用时，为了代码的模块化，强烈建议通过资源定义定时器。第二种方式普遍用于驱动中，因为在编写驱动的时候，并没有对应的资源文件。但不管是采用何种定义方式，定时器的使用方法及效果完全是一模一样的。

定时器的操作包括启动和停止，它有两套函数接口，这两套函数接口不可以交叉使用。一套接口需指定定时器 id，启动定时器的接口为：void StartTimer(U16 timerid, U32 delay, FuncPtr funcPtr)或者 void StartTimerEx(U16 timerid, U32 delay, oslTimerFuncPtr funcPtr, void* arg)，这两个函数的第一个参数是定时器 id，第二个参数是定时器触发的时间，单位为毫秒(1 秒=1000 毫秒)，第三个参数是定时器执行的函数。定时器从启动的时候开始计时，当达到触发时间时，就立即执行第三个参数传入的函数。StartTimerEx 还有第四个参数，这个参数是携带的用户数据，当定时器触发的时候，会把这个数据作为参数传入执行函数中。在开发的过程中，我们使用 StartTimer 比较多。以上两个函数启动的定时器，都必须使用 void StopTimer(U16 timerid)来停止。

定时器的另一套接口不需要指定定时器 id，启动定时器的接口为 void (*gui_start_timer) (S32 count, void(*callback)(void))或者 void(*gui_start_timer_ex) (S32 count, void (*callback) (void *), void *arg)。从其定义的原型可以看出，这是两个函数指针，它们分别被赋值为

UI_start_timer 和 UI_start_timer_ex，其参数列表中除了不需要定时器 id 外，其他的参数个数及含义分别与 StartTimer 和 StartTimerEx 对应。通常我们也是使用 gui_start_timer 比较多。而停止定时器，需要使用 void gui_cancel_timer(void (*callback) (void))。

虽然第二套定时器接口不需要我们传入定时器 id，但实际上这些 id 已被封装在函数体内部，当我们启动定时器的时候，由函数自动分配闲置的定时器 id，这个定时器 id 的取值范围为 UI_TIMER_ID_BASE~ UI_TIMER_ID_MAX，它们被定义在 TimerEvents.h 文件的 MMI_TIMER_IDS 枚举中。从代码中可以看出，默认的定时器 id 只有 10 个，我们可以加大这个范围，但需同时修改宏定义 MAX_UI_TIMERS 和枚举 UI_TIMER_ID_MAX 的值。在 MTK 系统中我们可以同时启动多个定时器，但是定时器比较耗资源，不能无限制地使用。

定时器启动一次，触发函数只会执行一次，如果我们要在定时器中重复执行某个操作，可在定时器的触发函数中再次启动它。还有一个用于判断定时器 id 的状态的函数接口：MMI_BOOL IsMyTimerExist(U16 nTimerId)，如果定时器执行结束或者已经通过 StopTimer 停止，这个函数都会返回 MMI_FALSE。

前面在讲解资源的章节中，我们已经添加了一个定时器资源 MYSTUDY_TIMER_ID。这一章节我们利用这个定时器来实现一个特效：不断地改变 IMG_STATIC_ID 图片的显示坐标，当它碰到屏幕边缘时再弹回来。

首先我们在进入 SCR_ID_HELLO_WORLD 屏幕后，使用 gui_start_timer 函数启动一个定时器，5 秒钟后清除屏幕上的所有内容，只显示图片 IMG_STATIC_ID，退出屏幕时，必须使用 gui_cancel_timer 停止定时器。在 MyStudyAppMain.c 文件中添加一个函数 mmi_mystudy_timer_handle，然后在 mtk_helloworld 函数中调用。代码如下(清单 4.1-1)：

--代码清单 4.1-1--

```
void mmi_mystudy_timer_handle(void)
{
    static S32 img_x=0, img_y=0, img_w=0, img_h=0;

    gdi_layer_clear(GDI_COLOR_BLACK);        /*设置背景色为黑色*/
    if(0==img_w || 0==img_h)                 /*初始化图片显示的起始位置和图片尺寸*/
    {
        gdi_image_get_dimension_id(IMG_STATIC_ID, &img_w, &img_h);
        img_x = (UI_DEVICE_WIDTH-img_w)/2;
        img_y = 0;
    }
    if(NULL != g_anim_handle)
    {
        gdi_anim_stop(g_anim_handle);        /*停止动画*/
        g_anim_handle = NULL;
    }
    gdi_image_draw_id(img_x, img_y, IMG_STATIC_ID);   /*显示图片*/
    gui_BLT_double_buffer(0, 0, UI_DEVICE_WIDTH, UI_DEVICE_HEIGHT);
```

```
    }
    void mtk_helloworld_exit(void)
    {
        gdi_anim_stop(g_anim_handle);/*停止动画*/
        g_anim_handle = NULL;
        gui_cancel_timer(mmi_mystudy_timer_handle);
    }
    void mtk_helloworld(void)
    {
        /*省略部分代码*/
        gui_start_timer(5000, mmi_mystudy_timer_handle);
        /*5 秒钟后执行 mmi_mystudy_timer_handle 函数*/
        SetRightSoftkeyFunction(mmi_frm_scrn_close_active_id, KEY_EVENT_UP);
    }
```

运行模拟器，进入 SCR_ID_HELLO_WORLD 屏幕，5 秒钟后就会进入如图 4.1-1 所示界面。

图 4.1-1

接下来，我们使用 StartTimer 启动自己定义的定时器资源 MYSTUDY_TIMER_ID，让图片一直不断地移动，时间间隔为 0.5 秒(500 毫秒)。修改 mmi_mystudy_timer_handle 函数，代码如下(清单 4.1-2)：

---代码清单 4.1-2---
```
    void mmi_mystudy_timer_handle(void)
    {
        static S32 img_x=0, img_y=0, img_w=0, img_h=0;
        static S16 x_flag=10, y_flag=10;

        /*省略部分代码*/
        if(img_x <= 0)
        {
```

```
            x_flag = 10;
        }
        else if(img_x+img_w >= UI_DEVICE_WIDTH)
        {
            x_flag = -10;
        }

        if(img_y <= 0)
        {
            y_flag = 10;
        }
        else if(img_y+img_h >= UI_DEVICE_HEIGHT)
        {
            y_flag = -10;
        }

        img_x = img_x+x_flag;
        img_y = img_y+y_flag;
        gdi_image_draw_id(img_x, img_y, IMG_STATIC_ID);/*显示图片*/

        gui_BLT_double_buffer(0, 0, UI_DEVICE_WIDTH, UI_DEVICE_HEIGHT);

        srv_backlight_turn_on(SRV_BACKLIGHT_SHORT_TIME);   /*禁止屏幕休眠*/
        StartTimer(MYSTUDY_TIMER_ID, 500, mmi_mystudy_timer_handle);/*启动定时器*/
}
```

再次运行模拟器，查看效果如图 4.1-2 所示。因为这个界面是动态的，图片会一直在屏幕上不断地跑，而书中只能展示静态的效果图。

以上代码的逻辑很简单，就是通过定时器每隔 500 毫秒调用一次 mmi_mystudy_timer_handle 函数，然后不断地改变图片显示的坐标，但是又不能让图片显示到屏幕之外的地方。代码中用到了一个陌生的函数 srv_backlight_turn_on(SRV_BACKLIGHT_SHORT_TIME)，这个语句的功能是防止屏幕进入休眠状态，如果屏幕休眠了，我们就看不到屏幕上的任何显示效果。读者可注释掉这条语句试一下效果，过一会屏幕就会变黑，但按任意键都会再次点亮屏幕。另外需要注意是，在退出屏幕的时候要把定时器停止，否则它会一直执行下去，所以还要修改 mtk_helloworld_exit 函数，代码如下(清单 4.1-3)：

图 4.1-2

---代码清单 4.1-3---

```
void mtk_helloworld_exit(void)
{
    /*省略部分代码*/
    if(MMI_TRUE==IsMyTimerExist(MYSTUDY_TIMER_ID))/*判断定时器状态*/
    {
        StopTimer(MYSTUDY_TIMER_ID);/*停止定时器*/
    }
}
```

　　定时器在 MTK 中是一个比较重要的内容,它的应用技巧也比较简单,分为单次调用和重复调用。以上代码中两种方式都展示了。至于两套定时器的接口如何选择,可根据实际情况而定。如果需要定时器实现的功能比较简单,并且使用的地方也比较少,可以选择 gui_start_timer,以省去定时器资源的创建;但如果定时器应用比较频繁,并且实现的功能也比较复杂,有时候需要跨越多个文件,则建议使用 StartTimer,毕竟有一个定时器 id 可以更方便在代码中控制它。以上实现的功能中,两套函数接口完全可以互换,实现的效果没有任何区别,读者可自己尝试。

4.1.2　Reminder 定时器

　　如果程序的执行频率只需要精确到分钟,那么用 Reminder 是最好的选择。它与系统时间的更新同步,不仅准时,而且比定时器节省资源。使用 Reminder 作为定时器,比定时器稍微复杂点,但也不是太难,总共分为五步。

　　(1) 定义 Reminder 消息 id。

　　在 plutommi\Service\ReminderSrv\ReminderSrvTypeTable.h 文件的 srv_reminder_type 枚举中修改代码,如代码清单 4.1-4 所示。

---代码清单 4.1-4---

```
typedef enum
{
        /*省略部分代码*/
#if defined(__MYSTUDY_APPLICATION__)
    SRV_REMINDER_TYPE_MYSTUDY,/*新增一个 reminder id*/
#endif
    /* Add reminder type before this line */
    SRV_REMINDER_TYPE_CUSTOM,                  /* 15 */
    SRV_REMINDER_TYPE_POWER_ON_CONFIRM,  /* 16 For power on confirm after alarm power on */
    SRV_REMINDER_TYPE_MRE_ALARM,
    SRV_REMINDER_TYPE_TOTAL                    /* 17 */
```

}srv_reminder_type;

我们新增了一个 Reminder id，命名为 SRV_REMINDER_TYPE_MYSTUDY，并用模块宏__MYSTUDY_APPLICATION__包含起来。这个枚举中有多个 Reminder id，新增的时候，一定要添加到 SRV_REMINDER_TYPE_CUSTOM 的前面。

(2) 添加 Reminder 消息资源。

在 CustomerAppMain.res 文件中修改代码如，代码清单 4.1-5 所示。

---代码清单 4.1-5---

```
/*省略部分代码*/
<!----------------消息资源-------------------------------------->
<SENDER id="EVT_ID_SRV_MYSTUDY_MSG_IND" hfile="MyStudyAppMain.h"/>
<RECEIVER id="EVT_ID_SRV_MYSTUDY_MSG_IND" proc="mmi_mystudy_msg_proc"/>
<RECEIVER id="EVT_ID_SRV_REMINDER_NOTIFY" proc="mmi_mystudy_reminder_proc"/>


<!----------------屏幕资源-------------------------------------->
/*省略部分代码*/
```

只要修改了资源文件，都需要使用 make resgen 指令编译资源。在讲解消息资源的时候已经介绍过，一个消息资源 id，可以有多个消息处理函数。读者可在 plutommi\Customer\CustomerInc\ mmi_rp_callback_mgr_config.h 文件中搜索 EVT_ID_SRV_REMINDER_NOTIFY，从 MMI_FRM_CB_REG_BEGIN(EVT_ID_SRV_REMINDER_NOTIFY)到 MMI_FRM_CB_REG_END(EVT_ID_SRV_REMINDER_NOTIFY) 之间所有的 MMI_FRM_CB_REG 里面指定的函数都是这个消息 id 的处理函数。上述资源中添加的 mmi_mystudy_reminder_proc 处理函数，编译资源后也应该出现在当中。当这个消息 id 发送消息时，这些函数都会被触发，此时就要在消息处理函数中根据第一步定义的 Reminder id 来区分哪个消息是我们需要的。

(3) 定义 Reminder 消息处理函数。

上一步资源中添加的消息处理函数并没有被定义，我们需要定义并实现它。在 MyStudyAppMain.c 文件中定义 mmi_mystudy_reminder_proc 函数，如代码清单 4.1-6 所示。

---代码清单 4.1-6---

```c
#include "ReminderSrvGprot.h"
/*省略部分代码*/
mmi_ret    mmi_mystudy_reminder_proc(mmi_event_struct *evt)
{
  srv_reminder_evt_struct *reminder_evt = (srv_reminder_evt_struct*)evt;

  if (reminder_evt->reminder_type!=SRV_REMINDER_TYPE_MYSTUDY
    && reminder_evt->reminder_type!=SRV_REMINDER_TYPE_TOTAL)
  {
```

```
                return MMI_RET_OK;
            }

            switch (reminder_evt->notify)
            {
                case SRV_REMINDER_NOTIFY_INIT:/*初始化*/
                    break;
                case SRV_REMINDER_NOTIFY_EXPIRY:/*触发*/
                    break;
                case SRV_REMINDER_NOTIFY_REINIT:/*更新*/
                    break;
                case SRV_REMINDER_NOTIFY_DEINIT:/*销毁*/
                    break;            }
            }
            return MMI_RET_OK;
        }
    #endif
```

上述函数只是实现了一个框架。函数开头定义了一个 Reminder 消息的结构体指针，并把参数传过来的指针进行强制类型转换。接下来的 if 条件判断就是为了过滤消息，这里不仅需要识别我们自己的 Reminder id，还需要识别系统的，因为系统每次更新时间都会触发这个函数，但并不是发送 reminder id SRV_REMINDER_TYPE_MYSTUDY。接收到消息数据后，需要处理四个 switch case 分支：每次 RTC 上电启动的时候都会执行 SRV_REMINDER_NOTIFY_INIT 分支，这里主要处理初始化的操作；当触发条件满足时就会执行 SRV_REMINDER_NOTIFY_EXPIRY 分支，我们把定时器需要处理的代码逻辑都加在这里执行；系统时钟每次更新的时候都会执行 SRV_REMINDER_NOTIFY_REINIT 分支，但系统时间被修改后，需要在这里重新计算定时器触发的时间；当把定时器销毁，就会执行 SRV_REMINDER_NOTIFY_DEINIT，通常这里做一些资源释放的工作。

（4）启动 Reminder。

使用 Reminder 做定时器，无法指定延时多长时间后执行某个操作，只能指定某个具体的时间。只有设置了 Reminder 的响应时间，并且响应时间大于当前时间，Reminder 定时器才算被启动了。在 MyStudyAppMain.c 文件中添加一个函数：mmi_set_reminder_time，代码如下（代码清单 4.1-7）。

---代码清单 4.1-7---
```
#include "DateTimeGprot.h"
/*省略部分代码*/
void mmi_set_reminder_time(void)
{
    MYTIME reminder_time={0x00};
```

```
        U32 datetime_sec = 0;

        GetDateTime(&reminder_time);/*获取系统当前时间*/
        datetime_sec = mmi_dt_mytime_2_utc_sec(&reminder_time, MMI_FALSE);/*将时间转换成
时间戳*/
        mmi_dt_utc_sec_2_mytime(datetime_sec+60, &reminder_time, MMI_FALSE);/*时间戳加60
秒，再转换成时间*/
        srv_reminder_notify_finish(MMI_FALSE);/*设置reminder提醒结束状态为 false*/
        srv_reminder_cancel(SRV_REMINDER_TYPE_MYSTUDY, 0);
        srv_reminder_set(SRV_REMINDER_TYPE_MYSTUDY, &reminder_time, 0); /*设置定时器
响应时间*/
    }
```

设置响应时间需要调用函数接口 srv_reminder_ret_enum srv_reminder_set(srv_reminder_type type, const MYTIME *expiry_time, U32 usr_data)，函数的第一个参数是我们定义的 Reminder id；第二个参数是定时器响应的时间；第三个参数可以不用，我们传入 0 就可以了。如果设置成功，会返回 SRV_REMINDER_RET_OK。

上述函数中，我们使用了时间相关的函数接口，这些函数可在 DateTimeType.h 文件中查看。其中涉及一个"时间戳"的概念，"时间戳"是指格林威治时间(0时区)1970 年 01 月 01 日 00 时 00 分 00 秒起至现在的总秒数，这是一个专业词汇，读者可上网查找。

在获取定时器响应时间时，先获取系统当前时间，然后转换成时间戳加60秒，再转换成时间，得到的定时器响应时间就比当前时间多一分钟。这里为什么要使用时间戳而不是直接使用时间的分钟数加 1 呢？如果当前时间是 59 分，加 1 就变成了 60，这肯定是不对的，而且我们还需要进行时间的进制转换，年、月、日、时都需要重新计算。而转换成时间戳后，时间的单位被统一成秒了，这样的运算就简便很多。

把上面的函数放到 mtk_helloworld 函数中调用，并注释掉 gui_start_timer(5000, mmi_mystudy_timer_handle)语句。然后在 mmi_mystudy_reminder_proc 函数的 SRV_REMINDER_NOTIFY_EXPIRY 分支中添加代码，具体见代码清单4.1-8。

---代码清单4.1-8---
```
mmi_ret mmi_mystudy_reminder_proc(mmi_event_struct *evt)
{
        srv_reminder_evt_struct *reminder_evt = (srv_reminder_evt_struct*)evt;
        MYTIME curr_time={0x00};
        U8 dt_str[64] = {0x00};
        /*省略部分代码*/
        switch (reminder_evt->notify)
        {
            /*省略部分代码*/
            case SRV_REMINDER_NOTIFY_EXPIRY:/*触发*/
```

```
            {
                GetDateTime(&curr_time);
                kal_wsprintf((WCHAR*)dt_str, "%04d/%02d/%02d %02d/%02d", curr_time.nYear,
                curr_time.nMonth, curr_time.nDay, curr_time.nHour, curr_time.nMin);
                gui_move_text_cursor(10, UI_DEVICE_HEIGHT-30);
                gui_print_text((UI_string_type)dt_str);/*在屏幕上打印字符*/
                gui_BLT_double_buffer(10, UI_DEVICE_HEIGHT-30, UI_DEVICE_WIDTH,
                UI_DEVICE_HEIGHT);
                break;
            }
            /*省略部分代码*/
    }
    return MMI_RET_OK;
}
```

以上代码实现的功能是：当 Reminder 触发后，就会在屏幕下方打印出当前时间，这里我们用到了一个 kal_wsprintf 函数，这个函数可以直接把数值格式化成 UCS2 编码格式的字符串，接下来在调用 gui_BLT_double_buffer 刷新屏幕的时候，我们只是局部刷新，运行代码一分钟后，模拟器显示效果如图 4.1-3 所示。

图 4.1-3

如果我们要按一定的频率重复执行某个操作，那该怎么处理呢？其实逻辑与普通定时器类似，即每次定时器触发后，再次设置定时器的触发时间为下一分钟。现在我们来实现一个电子手表的功能，让界面上的时间每隔一分钟更新一次，并且在系统时间被改变的时候也要重新更新定时器触发时间。修改代码如下(代码清单 4.1-9)：

--代码清单 4.1-9--
```
mmi_ret mmi_mystudy_reminder_proc(mmi_event_struct *evt)
{
    srv_reminder_evt_struct *reminder_evt = (srv_reminder_evt_struct*)evt;
    MYTIME curr_time={0x00};
    U8 dt_str[64] = {0x00};
    /*省略部分代码*/
    switch (reminder_evt->notify)
    {        /*省略部分代码*/
            case SRV_REMINDER_NOTIFY_EXPIRY:/*触发*/
            {
                GetDateTime(&curr_time);
```

```
            gui_fill_rectangle(10, UI_DEVICE_HEIGHT-30, UI_DEVICE_WIDTH, UI_DEVICE_
                HEIGHT , UI_COLOR_BLACK);        /*用背景色刷掉上一次显示的时间*/
            /*省略部分代码*/
            gui_BLT_double_buffer(10, UI_DEVICE_HEIGHT-30, UI_DEVICE_WIDTH,
                UI_DEVICE_HEIGHT);
            mmi_set_reminder_time();            /*设置响应时间为下一分钟*/
            break;
        }
        case SRV_REMINDER_NOTIFY_REINIT:        /*更新*/
        {
            mmi_set_reminder_time();    /*当系统时间被改变时,更新响应时间为下一分钟*/
            break;
        }
        /*省略部分代码*/
    }
    return MMI_RET_OK;
}
```

每次更新显示时间的时候,都必须把上一次显示的时间用背景色覆盖清空,否则显示会重叠。这里使用 gui_fill_rectangle 函数,也是局部填充。请读者自己运行看看效果。

(5) 停止 Reminder。

跟普通定时器一样,不用的时候我们要停止。因为 Reminder 定时器使用的是 RTC 中断,而 RTC 在系统关机的状态下并不会停止,否则系统时间就会丢失,除非完全断电。所以 Reminder 在关机的时候要停止,在 MyStudyAppMain.c 文件中添加 mmi_reset_reminder_time 函数用于停止 Reminder。详见代码清单 4.1-10。

--代码清单 4.1-10--
```
void mmi_reset_reminder_time(void)
{
    srv_reminder_notify_finish(MMI_TRUE);/*设置 reminder 消息提醒状态为 true*/
    srv_reminder_cancel(SRV_REMINDER_TYPE_MYSTUDY, 0);/*取消 reminder 消息*/
}
```

停止 Reminder 使用函数接口 MMI_BOOL srv_reminder_cancel(srv_reminder_type type, U32 usr_data),把函数 mmi_reset_reminder_time 放到 mtk_helloworld_exit 函数中调用,在退出 SCR_ID_HELLO_WORLD 屏幕的时候停止它。

4.2 层

在频繁更新的界面中,如果某些元素一直没有变化,我们就可以将这些元素提取出来画到一个模拟的屏幕中，而将一些需要更新的元素画到另外的模拟屏幕中，而后将两个模拟屏幕合并到真正的屏幕上，这样我们就节省了不变元素的重画时间，从而减轻了系统负担并加速画面更新。我们把这样的模拟屏幕叫做层，也可以说层就是屏幕的缓冲空间。实际上屏幕本身也是一个层，这个层不需要程序员创建，可以直接使用，我们称之为基础层，不管程序中创建了多少个层，最终都要合并到基础层上，才能正常显示。

层在 MTK 界面开发中主要用于实现各种特效，比如菜单滑动、图片透明等效果，实际上动画的播放也是通过层来实现的。MTK 系统中关于层的接口基本上都集中在 gdi_layer.c 文件中。其使用步骤如下：

1．定义层的句柄

层的句柄就是一个 gdi_handle 类型的变量，每个层都应该有一个句柄，它就相当于层的一个指针，或者通俗点理解为层的名字。我们对层的任何操作，都是通过句柄来实现的。因为对一个层的操作可能涉及多个函数甚至多个文件，为了代码结构的紧凑性，我们应该尽量避免一个层跨多个文件操作的情况，所以我们通常将句柄定义为静态全局变量，并设置初始值为 GDI_NULL_HANDLE。当层的句柄值为 GDI_NULL_HANDLE 时，表示该句柄不指向任何一个层。

定义层的句柄示例语句：static gdi_handle g_layer_handle = GDI_NULL_HANDLE;

2．创建层

创建层最常用的函数为 gdi_layer_create(OFFSET_X, OFFSET_Y, WIDTH, HEIGHT, HANDLE_PTR)，该函数有五个参数：OFFSET_X, OFFSET_Y 是层相对于屏幕原点的位置坐标；WIDTH, HEIGHT 是层的大小；HANDLE_PTR 就是我们前面创建的层句柄指针，用于保存所创建层的句柄。这里需要注意的是，使用 gdi_layer_create 函数创建的所有层，使用的都是系统预留的内存空间，其大小总和通常不能超过一个屏幕大小，也就是不能超过 LCD_WIDTH*LCD_HEIGHT。如果需要创建多个与屏幕尺寸大小相同的层，就需要用到函数 gdi_layer_create_using_outside_memory(X, Y, WIDTH, HEIGHT, HANDLE_PTR, OUTMEM_PTR, OUTMEM_SIZE)，这个函数前五个参数和 gdi_layer_create 的参数相同，而 OUTMEM_PTR 和 OUTMEM_SIZE 分别是存放层的内存空间和内存大小的，这个内存可以由程序员自己分配。

3．激活层

在 MTK 的图形系统中，任何时刻有且只能有一个层处于激活状态，所有的绘画都会默认画到当前激活层中，所以想要在层上绘画必须先将其激活。激活层的函数是 gdi_layer_set_active(gdi_handle handle)，handle 是创建层时的句柄。不过，由于在多层的处理中需要在各个层之间切换激活，所以我们经常用到的是 gdi_layer_push_and_set_active(gdi_handle handle)，此函数会把当前的激活层入栈而激活参数层，等到下次需要激活栈中的层时，只需要用函数 gdi_layer_pop_and_restore_active()激活就可以了。

4．绘制层上的内容

激活层之后，就可以在层上画任何东西了，包括字符、图片、动画等。不过此时的绘画坐标，是相对于层的坐标，而不是相对于屏幕的坐标。比如层在屏幕中的位置为(10, 10)，我们在层上显示一个图片，显示坐标为(10, 10)，则该图片显示的位置换算成相对于屏幕的坐标为(20, 20)。

5．合并到基础层

系统开机的时候会为每个硬件屏幕创建一个基础层，基础层有以下几个特点：

(1) 基础层由系统创建，无法删除。

(2) 与硬件屏幕完全重合。

(3) 系统默认的激活层，进入一个 Screen 时系统会自动将基础层激活。

(4) 显示更加快速，基础层存储于芯片内的 Flash 中，所以在其上面绘画极快，一般我们会将刷新频繁的内容放在基础层上。

基于以上几点，通常在不使用多层的情况下，我们完全可以将基础层当成硬件屏幕来看待，也就是说普通程序完全可以忽略层的概念。另外，因为系统一般只在进入某个 Screen 时才会自动将基础层激活，为避免特殊情况下使用层混乱，通常在新层上绘画完毕后我们会主动将基础层还原为激活状态。其实，对于我们上面说到的层之间的切换激活而言，大多数情况下是基础层和新层之间的切换，为此需要用到 GDI_RESULT gdi_layer_get_base_handle (gdi_handle *handle_ptr)得到基础层句柄，这样我们就可以切换激活层了。有了多个层，当然要合并到一起。合并层的函数是 gdi_layer_blt_previous(S32 x1, S32 y1, S32 x2, S32 y2)，而函数 gui_BLT_double_buffer 也具有同样的效果。另外，在合并前，我们还需要用函数 gdi_layer_set_blt_layer(H1, H2, H3, H4)来指明需要合并的层。MTK 中最多支持六个层的合并，但也有些平台只支持四个层的合并。

6．释放层

由于层需要占用一定的内存空间，所以我们创建的新层在使用完后一定要释放，以防止不必要的内存消耗。层的释放函数是 gdi_layer_free(gdi_handle handle)。我们通常都是在屏幕退出函数中释放。

接下来，我们使用层来实现一个图片的半透明显示效果。在 MyStudyAppMain.c 文件中新增一个函数，名为 mtk_mystudy_layer，然后把这个函数放到 mtk_helloworld 函数中调用，如代码清单 4.2-1 所示。

---代码清单 4.2-1---

```
/*省略部分代码*/
static gdi_handle g_layer_handle = GDI_NULL_HANDLE;      /*层的句柄*/
void mtk_mystudy_layer(void)
{
    gdi_handle base_handle = GDI_NULL_HANDLE;

    gui_move_text_cursor(70, 220);                       /*移动字符显示坐标为*/
    gui_print_text((UI_string_type)GetString(STR_ID_HELLO_WORLD));/*在屏幕上打印字符*/

    if(GDI_NULL_HANDLE == g_layer_handle)
    {
```

```
        gdi_layer_create(70, 200, 100, 100, &g_layer_handle);        /*创建层*/
    }
        gdi_layer_push_and_set_active(g_layer_handle);                /*激活层*/
        gdi_layer_clear(GDI_COLOR_RED);                               /*设置层的背景色*/

        gdi_image_draw_resized_id(20, 20, 60, 60, IMG_STATIC_ID);   /*在层上显示图片*/

        gdi_layer_set_opacity(MMI_TRUE, 255); /*设置层的透明度,取值0～255,值越低,透明越明显*/
        gdi_layer_pop_and_restore_active();

        gdi_layer_get_base_handle(&base_handle);
        gdi_layer_set_blt_layer(base_handle, g_layer_handle, GDI_NULL_HANDLE, GDI_NULL_HANDLE);
        gdi_layer_blt_previous(0, 0, LCD_WIDTH, LCD_HEIGHT);
    }

    void mtk_helloworld_exit(void)
    {
        /*省略部分代码*/
        if(GDI_NULL_HANDLE != g_layer_handle)
        {
            gdi_layer_free(g_layer_handle);/*释放层*/
            g_layer_handle = GDI_NULL_HANDLE;
        }
    }

    void mtk_helloworld(void)
    {
        /*省略部分代码*/
        mmi_mystudy_audio_play();
        mtk_mystudy_layer();
        /*省略部分代码*/
    }
```

上述代码在 mtk_mystudy_layer 函数中,我们首先打印了一行字符,这行字符显示在基础层上;然后新建一个层,并在层上绘制一张图片;最后在退出屏幕时,在屏幕退出函数 mtk_helloworld_exit 中把层释放掉。运行模拟器,显示效果如图 4.2-1 所示。

我们打印字符的坐标为(70, 220),而层的坐标为(70, 200),所以字符和层是重叠的,字符被层的内容覆盖掉了。如果我们把层设置为半透明状态,是不是就能透过层看到下面的字符内容呢？在代码清单 4.2-1 中，把语句"gdi_layer_set_opacity(MMI_TRUE, 255);"中的 255 改为 128。再次运行模拟器，显示效果如图 4.2-2 所示。

图 4.2-1

图 4.2-2

函数 gdi_layer_set_opacity 的功能就是设置层的透明度，第一个参数是一个布尔型，表示是否允许透明；第二个参数为透明度，取值范围为 0～255，值越大透明度越低，0 表示完全透明，255 表示完全不透明。如果第一个参数为 MMI_FALSE，则不允许透明，第二个参数就不起任何作用。

层的应用在 MTK 界面开发中属于一种高级技巧，通过它可以实现一些炫酷的效果，比如通过定时器不断改变层的坐标，可以实现悬浮窗的特效；还可以对层进行旋转，实现百叶窗的效果等。但是在智能穿戴和物联网设备研发中，层应用得比较少，因为这些设备的屏幕都比较小，显示的内容相对也简单些，一个基础层就够用了。不过也有一些智能手表上会开发一些互动性的小游戏，在 MTK 游戏开发中，层的应用就很频繁了。

4.3 文件管理

MTK 功能机平台最早是用于手机的操作系统，我们可以使用手机听音乐、看视频、阅读电子书，也可以拍照、录音、摄像等。这些丰富的功能都是基于文件系统实现的，实际上就是文件的读写操作。可以说，在任何平台上开发应用程序，文件的输入输出都是最基础的。凡是需要持久化存储的数据，都离不开文件，甚至我们可以使用文件系统来实现一个简单的数据库。

MTK 中的文件操作方法跟 C 语言的类似，只是函数接口的名称不一样。相关的函数接口都包含在头文件"interface\fs\Fs_gprot.h"中。文件管理可以分成两个部分：目录管理和文件操作。

4.3.1 目录管理

MTK 系统也有类似于 Windows 一样的磁盘空间，我们所有的文件目录，都要从磁盘的根目录开始访问。MTK 的模拟器中也实现了对磁盘空间的模拟，打开 MoDIS_VC9\WIN32FS 目录，里面有四个文件夹，如图 4.3-1 所示。

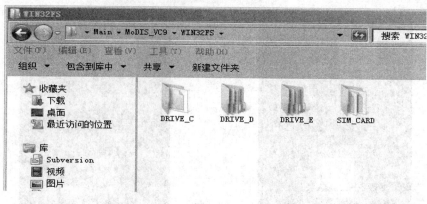

图 4.3-1

其中 SIM_CARD 在模拟器上相当于 SIM 卡，如果把这个文件夹删掉，运行模拟器就会在 idle 界面提示"请插入 SIM 卡"，相当于真机上没插入 SIM 卡的状态。SIM 卡中也有自己的文件系统，只不过空间比较小，我们一般都是往里面存短信和电话本，不会往里面存其他的一些用户数据。

另外三个文件夹在真机上相当于三个磁盘。"DRIVE_C"为"C"盘，也叫系统盘，它对用户是不可见的，主要用于存储系统数据，比如 NVRAM、电话本、短信等。在模拟器上，我们如果修改了 NVRAM 数据，需要格式化，则直接删除这个文件夹就行了。"DRIVE_D"的盘符为"D"，可称之为 D 盘，也叫手机盘或用户盘，它与 C 盘共享同一个存储空间，对用户是可见的，用 USB 数据线连接电脑，可以在电脑上显示该磁盘。不过并不是所有的 MTK 设备上都有 D 盘，它也取决于硬件的配置，一般先预留 C 盘的空间，如果还有剩余就用作 D 盘，没有剩余就不设 D 盘。"DRIVE_E"的盘符为"E"，是 MTK 设备上插入的 T 卡根目录，也属于用户盘。E 盘和 D 盘对用户来说是一样的，可以使用 E 盘(T 卡)来弥补没有 D 盘的不足。

在程序开发中，我们一般不推荐直接使用盘符访问，而是使用系统中的宏定义。比如系统盘(C 盘)的根目录可用 SRV_FMGR_SYSTEM_DRV 获取；手机盘(D 盘)的根目录可用 SRV_FMGR_PHONE_DRV 获取；T 卡(E 盘)的根目录可用 SRV_FMGR_CARD_DRV 获取。然而手机盘和 T 卡并不一定都存在，所以我们在开发应用程序的时候，要对此进行判断。比如：如果存在手机盘，则访问手机盘；如果不存在手机盘，则访问 T 卡；如果手机盘和 T 卡都不存在，则禁止运行应用程序。在 MTK 系统中还有一个获取磁盘根目录的宏定义，刚好能够满足这个要求，宏名称为 SRV_FMGR_PUBLIC_DRV，它默认访问的是手机盘，如果手机盘不存在，则访问 T 卡。在程序开发中，使用宏定义 SRV_FMGR_PUBLIC_DRV 来获取磁盘目录是比较常用的。

对文件目录的操作，最常用的就是创建和删除，创建目录的函数接口为 int FS_CreateDir(const WCHAR * DirName)，返回值为负数表示创建失败，参数"DirName"为文件夹路径，注意其类型为宽字符(WCHAR)。删除目录的函数接口为 int FS_RemoveDir(const WCHAR * DirName)，不过该函数只能删除空目录。接下来，我们通过代码创建一个目录，然后再把它删除。在 MyStudyAppMain.c 文件中新建函数 mtk_mystudy_filemanager，代码如代码清单 4.3-1 所示。

--代码清单 4.3-1--

```
void mtk_mystudy_filemanager(void)
{
    WCHAR path[64]={0x00};
    kal_wsprintf((WCHAR*)path, "%c:\\%w", SRV_FMGR_PUBLIC_DRV, L"myDir");
    FS_CreateDir(path);/*创建文件夹*/
    FS_RemoveDir(path); /*删除文件夹*/
}
```

以上代码很简单，我们先在手机盘内创建了一个文件夹"myDir"，然后又立即删除它。把函数放在 mtk_helloworld 中调用，然后运行模拟器，在 mtk_mystudy_filemanager 函数中打上断点，逐行执行函数中的每一条语句并查看 MoDIS_VC9\WIN32FS\DRIVE_D 目录下的变化就会发现，当执行完 FS_CreateDir(path)语句时，DRIVE_D 目录下会多出一个"myDir"文件夹，当执行完 FS_RemoveDir(path)语句时，"myDir"文件夹就消失了。

4.3.2 文件操作

MTK 系统中的文件操作，几乎与 C 语言中的文件操作是一模一样的，其函数名称也相似。我们对文件的操作主要包括创建、读写、删除。对文件进行读写之前必须先打开文件，而创建文件的动作就包含在打开之中。所以从代码层面上来看，我们对文件的操作主要包含打开文件、读写文件、删除文件。

1．打开文件

打开文件的函数接口为 int FS_Open(const WCHAR * FileName, kal_uint32 Flag)，其返回值为文件的句柄，有效值大于或等于 0，如果小于 0 则表示文件打开失败。第一个参数(FileName)为文件名称，需包含文件的路径；第二个参数(Flag)为文件的打开方式，常用的打开方式有只读(FS_READ_ONLY)、可读可写(FS_READ_WRITE)、创建(FS_CREATE)。如果是打开一个不存在的文件，则必须以创建的方式打开。

函数调用示例如下：

(1) FS_Open(L"C :\\mysutdy_file.txt ", FS_READ_WRITE);

以读写方式打开文件，如果文件不存在，则打开失败；如果文件存在，打开后既可以修改文件中的数据，也可以读取文件中的数据。

(2) FS_Open(L"C :\\mysutdy_file.txt ", FS_READ_WRITE|FS_CREATE);

以读写方式和创建方式打开文件，如果文件不存在，则先创建文件，再以可读可写的方式打开；如果文件已经存在，则只打开不创建。

(3) FS_Open(L"C :\\mysutdy_file.txt ", FS_READ_ONLY);

以只读方式打开文件，如果文件不存在，则打开失败；如果文件存在，打开后只能读取文件中的数据，无法对文件中的数据进行修改。

(4) FS_Open(L"C :\\mysutdy_file.txt ", FS_READ_ONLY |FS_CREATE);

以只读方式和创建方式打开文件,如果文件不存在,则先创建再以只读方式打开;如果文件已经存在,则以只读方式打开。

2. 读写文件

只有在文件打开的状态下,才能对它的内容进行读和写,而且文件的打开方式必须与操作对应,否则会报没有访问权限的错误。比如以只读模式打开一个文件,而我们却往里面写数据,这是不合法的。读文件的函数接口名称为 FS_Read,写文件的函数接口名称为 FS_Write,其函数说明分别如下:

(1) int FS_Read(FS_HANDLE FileHandle, void * DataPtr, kal_uint32 Length, kal_uint32 * Read)

函数的第一个参数(FileHandle)为文件的句柄,来源于 FS_Open 函数的返回值;第二个参数(DataPtr)是用于保存读取数据的内存指针;第三个参数(Length)是存储数据的内存大小,也就是第二个参数(DataPtr)指向的内存大小;第四个参数(Read)表示从文件中实际读出的数据字节个数,这个值一般都是小于或等于第三个参数(Length)值。函数的返回值小于 0 则表示读取失败,具体失败原因可参考 interface\fs\fs_gprot.h 文件中的枚举定义 fs_error_enum。

(2) int FS_Write(FS_HANDLE FileHandle, void * DataPtr, kal_uint32 Length, kal_uint32 * Written)

函数的第一个参数(FileHandle)是文件句柄,来源于 FS_Open 函数的返回值;第二个参数(DataPtr)中保存着待写入文件的数据;第三个参数(Length)表示待写入数据的字节大小;第四个参数(Written)表示实际写入文件中的数据字节个数,这个值一般情况下应等于第三个参数(Length)值。函数的返回值小于 0 则表示写文件失败,具体失败原因同样可参考 interface\fs \fs_gprot.h 文件中的枚举定义 fs_error_enum。

3. 关闭文件

如果我们对文件操作结束,就必须把文件关闭。关闭文件的函数接口为 int FS_Close(FS_HANDLE FileHandle),这个函数只有一个参数,只需要传入文件句柄就可以了,这个句柄同样来源于 FS_Open 函数的返回值。

接下来我们在 mtk_mystudy_filemanager 函数中添加一点内容,在用户盘中创建一个文件,写入一些数据,再读取出来,然后再关闭。代码如下(代码清单 4.3-2):

---代码清单 4.3-2---

```
void mtk_mystudy_filemanager(void)
{
    WCHAR path[64]={0x00}, file_name[64] = {0x00};
    FS_HANDLE file_handle = NULL;
    U32 write_size=0, read_size=0;
    U8 file_content[128]={0x00};

    kal_wsprintf((WCHAR*)path, "%c:\\%w", SRV_FMGR_PUBLIC_DRV, L"myDir");
```

```
FS_CreateDir(path);/*创建文件夹*/
FS_RemoveDir(path); /*删除文件夹*/


kal_wsprintf((WCHAR*)file_name, "%c:\\mysutdy_file.txt", SRV_FMGR_PUBLIC_DRV);
file_handle = FS_Open(file_name, FS_READ_WRITE|FS_CREATE);/*打开文件*/


sprintf((char*)file_content, "mtk mystudy filemanager");
FS_Write(file_handle, file_content, strlen(file_content), &write_size);/*写文件*/

memset(file_content, 0x00, sizeof(file_content));
FS_Read(file_handle, file_content, sizeof(file_content), &read_size);/*读文件*/

FS_Close(file_handle);/*关闭文件*/
}
```

运行模拟器，当执行完该函数后，会在"MoDIS_VC9\WIN32FS\DRIVE_D"目录下生成一个文件"mysutdy_file.txt"，文件中的内容为"mtk mystudy filemanager"，如图 4.3-2 所示。

图 4.3-2

但是上面的函数有一个问题，不管我们把函数执行多少次，文件中的内容依旧只有一个"mtk mystudy filemanager"字符串。从理论上说，我们每多执行一次函数，就会往文件中多写入一次数据，应该会出现多个"mtk mystudy filemanager"字符串才对，而且当我们使用断点跟踪时，会发现执行完 FS_Read 函数后，字符串 file_content 变量中的内容是空的，read_size 的值也为 0，这说明根本就没有从文件中读到数据。这是为什么呢？因为在文件中存在一个文件指针，每次打开文件的时候，文件指针都指向文件开头，每一次读写数据，文件指针都会移到最后一次读写的末尾处，所以我们每次往文件中写数据的时候都会覆盖掉上一次写的内容，而执行完 FS_Write 函数后，文件指针就指向了文件末尾处，此时再执行 FS_Read 根本就读不到任何内容，除非把文件指针重新指向文件开头处。修改 mtk_mystudy_filemanager 函数代码，在 FS_Read 函数前添加一行语句 FS_Seek(file_handle, 0, FS_FILE_BEGIN)，代码如下(代码清单 4.3-3)：

---代码清单 4.3-3---

```
void mtk_mystudy_filemanager(void)
{
    /*省略部分代码*/
    memset(file_content, 0x00, sizeof(file_content));
    FS_Seek(file_handle, 0, FS_FILE_BEGIN);/*移动文件指针，指向文件开头处*/
    FS_Read(file_handle, file_content, sizeof(file_content), &read_size);/*读文件*/

    FS_Close(file_handle);/*关闭文件*/
}
```

再次运行模拟器，使用断点跟踪，会发现执行完 FS_Read 语句后，file_content 变量中的内容就是我们前面写入的内容。

函数 int FS_Seek(FS_HANDLE FileHandle, int Offset, int Whence)的作用是移动文件指针，第一个参数(FileHandle)是操作文件的句柄；第二个参数(Offset)是指针移动的偏移量，取值小于 0 表示向前移动，等于 0 表示不移动，大于 0 表示向后移动；第三个参数(Whence)是指针移动的参考位置，这个变量的取值见 fs_gprot.h 文件中的枚举定义 FS_SEEK_POS_ENUM，其中"FS_FILE_BEGIN"是从文件起初处移动，只能往后移，此时 Offset 小于 0 是没有意义的，等于 0 则表示指针指向文件起始处；"FS_FILE_CURRENT"是从当前指针位置处开始移动，可往前移动也可往后移动；"FS_FILE_END"表示从文件末尾处开始移动，只能往前移，此时 Offset 大于 0 是没有意义的，等于则表示指针指向文件末尾处。

根据以上描述，如果我们每次往文件中写数据，都要在文件末尾处追加而不是覆盖，则只需要在打开文件后，开始写数据之前添加一行语句：FS_Seek(file_handle, 0, FS_FILE_END);让文件指针指向文件末尾处就可以了。

4.4 NVRAM

NVRAM(Non-Volatile Random Access Memory)是非易失性随机访问存储器，指断电后仍能保持数据的一种 RAM(Random Access Memory 随机存储器)，通常我们就简称为 NV。在 MTK 设备中，只要是断电后仍不能丢失的数据，都建议存储在 NV 中。

NV 有两种形式：一种只能保存简单的数值，这种 NV 我们通常使用资源的方式加载，它最大只能存储 8 个字节的数据；另一种存储的数据量比较大，通常用于存储复合数据，比如结构体，它在源文件中定义，只要 NVRAM 的内存足够，几乎可以保存任意大小的数据。我们也可以把简单数据使用第二种 NV 保存，两种 NV 在定义上和使用上都有差别，使用的函数接口也不一样，但不管哪种 NV，其操作方式只有读和写，函数接口都定义在

plutommi\Framework\NVRAMManager\NVRAMManagerSrc\NvramInterface.c 和 nvram\src\nvram_interface.c 文件中。

4.4.1 存储简单数据的 NVRAM

在前面讲解资源的时候，我们已经新增了一个 NV 资源：NVRAM_MYSTUDY_ID。这个 NV 资源的 Type 属性为"byte"，所以它只能存储一个字节的数据，对应的变量类型为 S8 和 U8。在 MyStudyAppMain.c 文件中新增一个函数，命名为 mmi_mystudy_nvram，并把它放到 mmi_highlight_mystudy 函数中调用。我们实现一个简单的功能，用 NV 记录 MAIN_MENU_MYSTUDY_ID 菜单的高亮次数。代码如下(代码清单 4.4-1)：

---代码清单 4.4-1---

```
void mmi_mystudy_nvram(void)
{
    S16 error = 0;
    U8 simple_nvram = 0;

    ReadValue(NVRAM_MYSTUDY_ID, &simple_nvram, DS_BYTE, &error);/*读 NV*/
    simple_nvram += 1;
    WriteValue(NVRAM_MYSTUDY_ID, &simple_nvram, DS_BYTE, &error); /*写 NV*/
}

void mmi_highlight_mystudy(void)
{
    /*省略部分代码*/
    mmi_mystudy_nvram();
    SetKeyHandler(mmi_entry_mystudy, KEY_LSK, KEY_EVENT_UP);
}
```

上述代码实现的逻辑很简单，每次执行 mmi_mystudy_nvram 函数时，先用函数 S32 ReadValue(U16 nDataItemId, void *pBuffer, U8 nDataType, S16 *pError) 把 NVRAM_MYSTUDY_ID 中的数据读取出来，然后加 1，再用函数 S32 WriteValue(U16 nDataItemId, void *pBuffer, U8 nDataType, S16 *pError)把数值写进 NVRAM_MYSTUDY_ID 中。第一次读取的时候，数值为 NV 的默认值，在添加资源时我们设置了默认值为 0。这两个函数的参数含义都是一样的，第一个参数(nDataItemId)是资源文件中定义的 NVRAM 资源 id；第二个参数(pBuffer)是存储数据的变量，只不过一个是把值传出，一个是把值传入；第三个参数(nDataType)表示 NV 中存储的数据类型，取值见 plutommi\Framework\CommonFiles\CommonInc\mmi_frm_nvram_gprot.h 文件中的枚举 DATASIZE，其中 DS_BYTE、DS_SHORT、DS_DOUBLE 分别对应 NV 资源的 Type 属性 byte、short、double；第四个参数(pError)暂时没用，但不能传入空指针，否则会死机，这是 MTK 原始代码上留的 bug，我们也可以修改

函数代码，加上指针是否为空的判断条件。

这次我们使用 VS2008 打断点的方式来查看 NV 中读出的数值。首先以调试模式(快捷键 F5)运行模拟器，在 VS2008 中按下快捷键 Ctrl+B，在弹出的新建断点提示框中输入函数名 mmi_mystudy_nvram，如图 4.4-1 所示，然后点击"确定"按钮，在断点窗口中就能看到已经打上的断点，如图 4.4-2 所示。

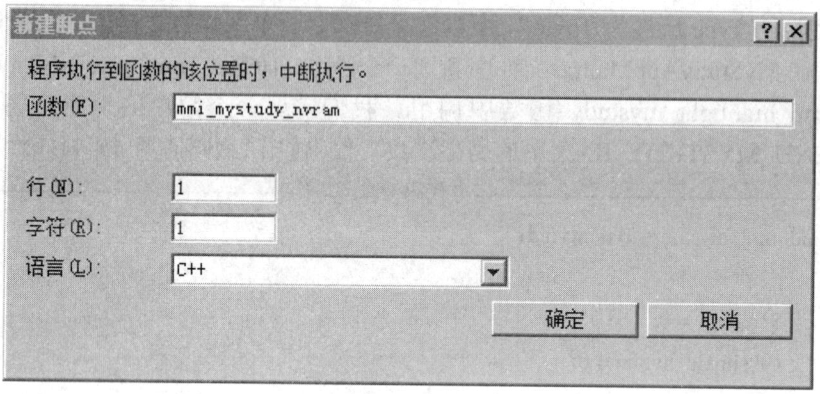

图 4.4-1

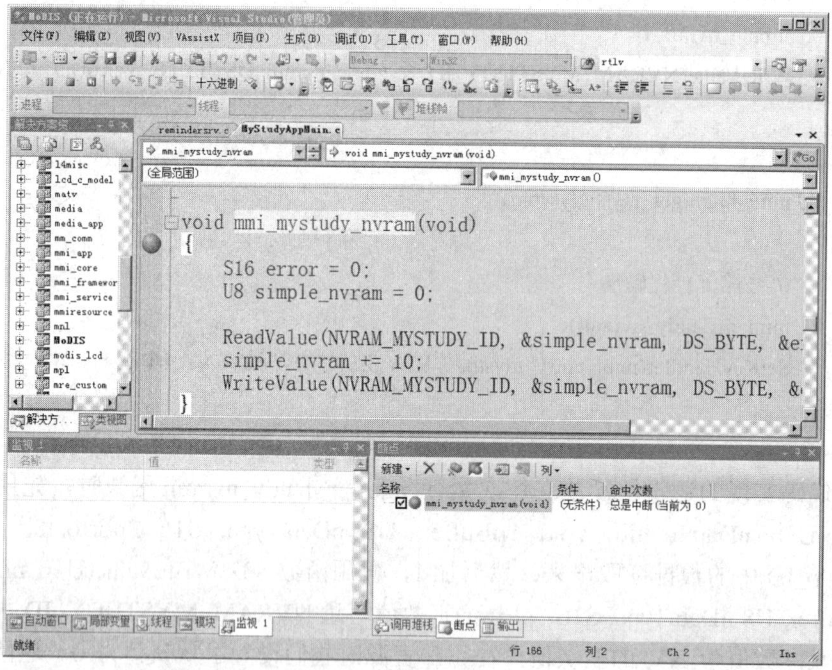

图 4.4-2

进入模拟器的主菜单界面，移动光标，每次选中 MAIN_MENU_MYSTUDY_ID 菜单的时候，都会执行 mmi_mystudy_nvram 函数，按快捷键 F10 逐行执行代码,当执行完 ReadValue 语句时，NV 中的数据就已经读取出来并存储到了变量 simple_nvram 中，我们双击选中 simple_nvram 变量，把它拖到监视窗口中，就能看到它在内存中的数值，如图 4.4-3 所示。继续按 F10，当执行完 WriteValue 语句时，NV 中保存的数据已经加 1 了。

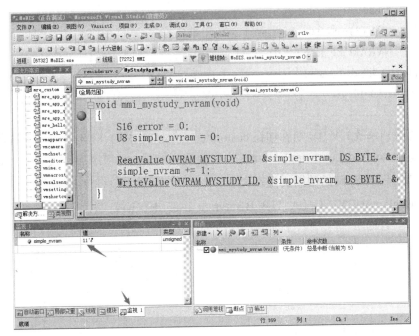

图 4.4-3

当图 4.4-3 中 MAIN_MENU_MYSTUDY_ID 菜单已经被高亮 11 次后，我们退出模拟器(Shift+F5)，结束调试，再次按 F5 以调试模式运行模拟器，相当于真机设备的关机重启操作，此时高亮选中 MAIN_MENU_MYSTUDY_ID 菜单，再次执行 mmi_mystudy_nvram 函数，会发现 NV 中读取出来的数据，依旧是我们上一次保存的数据。

4.4.2 存储复合数据的 NVRAM

添加存储复合数据的 NV，方法稍微复杂点，总共分为四步，其添加的地方也有两处，可以与系统中的 NV 添加到一起，文件目录为 custom\common\PLUTO_MMI；也可以按项目独立添加，文件目录为 custom\app\HEXING60A_11B_BB(注意：HEXING60A_11B_BB 是与项目相关的目录，会随项目名称而变化)。我们推荐使用后者，但按项目独立新增的 NV 的文件中存在一些 bug，这些 bug 是 MTK 原始代码中存在的，需要我们自己修改。

1. 新增 NVRAM id

在 custom\app\HEXING60A_11B_BB\nvram_user_defs.h 文件中的 nvram_lid_cust_enum 枚举中定义 NVRAM id，该文件中的代码如下(代码清单 4.4-2)：

--代码清单 4.4-2--

```
/*省略部分代码*/
typedef enum
{
    NVRAM_EF_PHONY_LID = NVRAM_LID_CUST_BEGIN,
#if defined(__MYSTUDY_APPLICATION__)
    NVRAM_EF_MYSTUDY_LID,  /*NVRAM ID*/
```

```
        #endif
        /* Don't remove this line: insert LID definition above */
        NVRAM_EF_LAST_LID_CUST
    } nvram_lid_cust_enum;
    /*省略部分代码*/
```

如果新增到系统 NV 中，则添加在 custom\common\PLUTO_MMI\nvram_common_defs.h 文件的枚举 nvram_lid_commapp_enum 中，需要注意的是，必须添加到最后一个 id 的上面。

2. 定义 NV 中存储的数据类型及大小

在 custom\app\HEXING60A_11B_BB\nvram_user_defs.h 文件中，添加代码如下(代码清单 4.4-3)：

--代码清单 4.4-3--

```
/*省略部分代码*/


/**********************************
 * Custom data item Define Start
 **********************************/
#if defined(__MYSTUDY_APPLICATION__)
typedef struct
{
    kal_uint16   highlight_count;
}stu_mystudy_nvram;           /*nv 中存储的数据类型*/

#define NVRAM_EF_MYSTUDY_SIZE      (sizeof(stu_mystudy_nvram))/*数据类型的大小*/
#define NVRAM_EF_MYSTUDY_TOTAL  1                             /*数据的个数*/
#endif


/* size 和 total 不能等于 0, 否则分配内存失败*/
#if defined(__MYSTUDY_APPLICATION__)
#define NVRAM_PHONY_SIZE      1
#define NVRAM_PHONY_TOTAL     1
#else
#define NVRAM_PHONY_SIZE      0
#define NVRAM_PHONY_TOTAL     0
#endif
/* Don't remove this line: insert size and total definition above */
/*省略部分代码*/
```

其中 NVRAM_EF_MYSTUDY_SIZE 表示数据类型的大小，NVRAM_EF_MYSTUDY_TOTAL 表示 NV 中总共需要存储的数据个数，NVRAM_EF_MYSTUDY_SIZE 乘以 NVRAM_EF_MYSTUDY_TOTAL 得到的值就是这个 NV 占用的总内存大小，这个大小必须为大于 0 的整数，所以这个文件中存在一个 bug，NVRAM_PHONY_SIZE 和 NVRAM_PHONY_TOTAL 都被定义为 0，这是不允许的，代码编译时会报错。在上述代码中我们修复这个 bug，将他们定义为 1。

如果定义在系统 NV 中，则数据类型(stu_mystudy_nvram)定义在 custom\common\PLUTO_MMI\common_nvram_editor_data_item.h 文件中，NV_SIZE(NVRAM_EF_MYSTUDY_SIZE) 和 NV_TOTAL(NVRAM_EF_MYSTUDY_TOTAL) 定义在 custom\common\PLUTO_MMI\nvram_common_defs.h 文件中。

3. 定义 NVRAM 的版本号

通常我们只要修改了 NVRAM，下载到真机设备后都必须进行格式化操作，让系统重新分配 NVRAM 的内存空间，否则会出现 NVRAM 访问出错而导致死机，但有时候我们虽然修改了 NVRAM，却不允许对系统进行格式化操作，此时只需把修改的 NVRAM 进行版本号升级，这个 NVRAM 的内存也会被重新分配并且清空之前的数据。NVRAM 的版本号命名必须是 NVRAM id 后面接"_VERNO"构成，其值必须是三个字符，默认从"000"开始，然后升级的时候依次递增。NVRAM 的版本号定义在 custom\app\HEXING60A_11B_BB\custom_nvram_editor_data_item.h 文件中，代码如下(代码清单 4.4-4)：

---代码清单 4.4-4---

```
/*省略部分代码*/
/* Version numbers of data items */
    #define NVRAM_EF_PHONY_LID_VERNO         "000"
#if defined(_MYSTUDY_APPLICATION_)/*定义 NVRAM 版本号*/
    #define NVRAM_EF_MYSTUDY_LID_VERNO       "000"
#endif
/* DO NOT MOVE OR REMOVE THIS HEADER */
#ifndef _OPTR_NONE_
#include "op_custom_nvram_editor_data_item.h"
#endif
/*省略部分代码*/
```

系统 NV 中的版本号定义在 custom\common\PLUTO_MMI\common_nvram_editor_data_item.h 文件中。

4. 填充 ltable_entry_struct 结构体数组

前面几步已经把 NVRMA 所需要的数据都定义了，最后一步要把这些数据整合到 custom\app\HEXING60A_11B_BB\nvram_user_config.c 文件的 ltable_entry_struct 结构体数组 logical_data_item_table_cust 中。这个文件中的原始代码也存在一些错误，我们一并把它修改好了，代码如下(代码清单 4.4-5)：

---代码清单 4.4-5---
```
/*省略部分代码*/

/*#if (NVRAM_EF_LAST_LID_CUST - NVRAM_LID_CUST_BEGIN > 1)  始终为 false*/
#if (NVRAM_EF_LAST_LID_CUST - NVRAM_LID_CUST_BEGIN > 1)||defined(_MYSTUDY_
    APPLICATION_)
ltable_entry_struct logical_data_item_table_cust[] =
{
    {
        NVRAM_EF_PHONY_LID,
        NVRAM_PHONY_TOTAL,
        NVRAM_PHONY_SIZE,
        NVRAM_NORMAL(NVRAM_EF_ZERO_DEFAULT),
        NVRAM_CATEGORY_USER,
        NVRAM_ATTR_AVERAGE,
        "CT00",
        VER(NVRAM_EF_PHONY_LID)
    },
#if defined(__MYSTUDY_APPLICATION__)/*定义 NVRAM 版本号*/
    {
        NVRAM_EF_MYSTUDY_LID,       /*NVRAM id*/
        NVRAM_EF_MYSTUDY_TOTAL,/*数据个数*/
        NVRAM_EF_MYSTUDY_SIZE,  /*数据类型大小*/
        NVRAM_NORMAL(NVRAM_EF_ZERO_DEFAULT), /*默认值*/
        NVRAM_CATEGORY_USER,      /*nvram 的种类*/
        NVRAM_ATTR_AVERAGE|NVRAM_ATTR_FACTORY_RESET, /*nvram 的属性*/
        "MST0", /*nvram 的标志*/
        VER(NVRAM_EF_MYSTUDY_LID)/*版本号*/
    },
#endif
    /* Watch out! There is no comma after last element! */
#if !defined(__MYSTUDY_APPLICATION__)/* 如果不去掉，nvram_auto_gen.exe 编译通不过*/
    { NVRAM_EF_RESERVED_LID }
#endif
};

#if defined(__MYSTUDY_APPLICATION__)/* ltable_entry_struct 编译报错*/
nvram_ltable_tbl_struct nvram_ltable_cust = {logical_data_item_table_cust, sizeof(logical_data_
    item_table_cust)/sizeof(nvram_ltable_entry_struct)};
```

```
#else
nvram_ltable_tbl_struct nvram_ltable_cust = {logical_data_item_table_cust, sizeof(logical_data_
item_table_cust)/sizeof(ltable_entry_struct)};
#endif
#endif
```

上述代码填充的结构体中，前三项在前面已经介绍过了，第四项表示NVRAM内存中的默认值，如果置为空的话通常使用NVRAM_EF_ZERO_DEFAULT(0x00)和NVRAM_EF_FF_DEFAULT(0xFF)，我们也可以自定义默认值；第五项表示NVRAM的种类，它的取值只有两种：NVRAM_CATEGORY_USER和NVRAM_CATEGORY_INTERNAL，NVRAM_CATEGORY_INTERNAL是系统内部的NVRAM，通常我们自己定义的NVRAM都是NVRAM_CATEGORY_USER类型；第六项表示NVRAM的属性，NVRAM_ATTR_AVERAGE表示普通属性，也可以当做默认属性，NVRAM_ATTR_FACTORY_RESET类似于NVRAM资源的restore_flag="TRUE"，恢复出厂设置时会把NV中的数据还原成默认值；第七项是NVRAM的标志，这个标志是一个占4字节的字符串，它必须在所有NVRAM中唯一，如果有重名，编译过程中也会报错；第八项是NVRAM的版本号，VER是一个宏定义，它的作用是在NVRAM id后面添加"_VERNO"字符，这就是为什么定义NVRAM版本号时一定要使用NVRAM id加"_VERNO"来命名的原因。

如果添加到系统NV中，则在custom\common\PLUTO_MMI \nvram_common_config.c文件的logical_data_item_table_common_app结构体数组中添加。

至此，一个存储复合数据的NVRAM就已经添加成功了，接下来我们可以对其进行读写操作。复合数据的NVRAM读写接口使用S32 ReadRecord(nvram_lid_enum nLID, U16 nRecordId, void *pBuffer, U16 nBufferSize, S16 *pError)和S32 WriteRecord(nvram_lid_enum nLID, U16 nRecordId, void *pBuffer, U16 nBufferSize, S16 *pError)，这两个函数的参数也是一样的，第一个参数(nLID)是我们定义的NVRAM id；第二个参数(nRecordId)表示NVRAM中的第几个数据记录，取值范围为1~NVRAM_TOTAL(NVRAM_EF_MYSTUDY_TOTAL)；第三个参数(pBuffer)存储数据的变量指针；第四个参数(nBufferSize)是NVRAM中存储的数据类型的大小(NVRAM_EF_MYSTUDY_SIZE)；第五个参数(pError)同样是无效的参数，但不能传入空指针，它总是被赋值为NVRAM_WRITE_SUCCESS。

修改mmi_mystudy_nvram函数，我们添加复合数据的NVRAM读写方式。代码如下(代码清单4.4-6)：

---代码清单4.4-6---
```
#include "nvram_user_defs.h"
void mmi_mystudy_nvram(void)
{
    S16 error = 0;
    U8 simple_nvram = 0;
    stu_mystudy_nvram nvram_date={0x00};
```

```
/*省略部分代码*/
ReadRecord(NVRAM_EF_MYSTUDY_LID, 1, &nvram_date, NVRAM_EF_MYSTUDY_SIZE, &error);
nvram_date.highlight_count += 1;
WriteRecord(NVRAM_EF_MYSTUDY_LID, 1, &nvram_date, NVRAM_EF_MYSTUDY_SIZE, &error);
}
```

请读者以同样的方式验证代码,但在运行模拟器前,先把 MoDIS_VC9\WIN32FS 目录下 DRIVE_C 文件夹删除掉,相当于真机设备的格式化操作。这个文件夹在有的系统中是被隐藏的,需先显示隐藏文件和文件夹。

针对存储复合数据的 NVRAM,在 MTK 系统中还有一组读写方式,这种读写方式可以应用到底层代码中,比如在执行 AT 指令的过程中,使用上述的方式读写 NVRAM 会导致死机,而使用这种读写方式不会死机。函数接口原型如下:

```
kal_bool  nvram_external_read_data(nvram_lid_enum LID, kal_uint16 rec_index, kal_uint8 *buffer, kal_uint32 buffer_size)

kal_bool  nvram_external_write_data(nvram_lid_enum LID, kal_uint16 rec_index, kal_uint8 *buffer, kal_uint32 buffer_size)
```

这两个函数跟以上两个函数的用法相同,只是少了最后一个参数,在 mmi_mystudy_nvram 函数中添加示例代码如下(代码清单 4.4-7):

---代码清单 4.4-7---
```
void mmi_mystudy_nvram(void)
{
    /*省略部分代码*/
    nvram_date.highlight_count = 0;
    nvram_external_read_data(NVRAM_EF_MYSTUDY_LID, 1, &nvram_date, sizeof(nvram_date));
    nvram_external_write_data(NVRAM_EF_MYSTUDY_LID, 1, &nvram_date, sizeof(nvram_date));
}
```

第五章 网络与定位

5.1 SIM 卡通信

MTK 系统最早主要应用于手机,大家都知道手机必须插入 SIM(Subscriber Identification Module)卡才能打电话、发短信,如果开通了流量功能,还可以用来上网。虽然现在大多数 MTK 产品都脱离了手机的形态,但 SIM 通信依旧是一个很重要的功能,在 MTK 功能机平台上的 SIM 通信通常也被称之为 GSM(Global System for Mobile Communication)通信。GSM 是第二代移动电话系统,也就是所谓的 2G,而基于 GSM 的上网功能被称之为 GPRS(General Packet Radio Service)。虽然现在已经发展到了 5G 时代,但大多数 MTK 平台依旧使用的是 2G 通信标准,也有部分平台使用的是 3G。今后随着行业的发展,相信 MTK 平台也会不断地提高通信标准。但不管它使用第几代通信标准,对于打电话、发短信、上网等功能在代码上的实现方式,几乎没有任何区别。

5.1.1 短信

"短信"英文名称为 Short Message Service,简称 SMS。使用过手机的人应该都知道它具有什么样的功能。在 MTK 系统中,有一套完整的短信接口,我们只需要了解怎么使用就可以了。

理论上,任何字符都可以通过短信发送,包括各国语言文字、数字等。在 MTK 系统中发送短信的接口也有多个,但我们常用的只有两个,函数原型如下所示,它们都定义在 plutommi\Service\SmsSrv\ SmsSenderSrv.c 文件中。

(1) void srv_sms_send_ucs2_text_msg(char *ucs2_content,
 char *ucs2_number, srv_sms_sim_enum sim_id,
 SrvSmsCallbackFunc callback_func, void *user_data)

(2) void srv_sms_send_asc_text_msg(char *asc_content, char *asc_number,
 srv_sms_sim_enum sim_id,
 SrvSmsCallbackFunc callback_func, void *user_data)

这两个函数的参数个数和含义都是一样的,使用方法也是一样的。第一个参数(ucs2_content、asc_content)是短信要发送的字符内容,两个函数的区别在于编码格式不同,前者使用 UCS2 编码,后者使用 ASC 编码;第二个参数(ucs2_number、asc_number)是接收短信的电话号码,同样是编码格式不同;第三个参数(sim_id)是发送短信的 SIM 卡,如果只有一个 SIM,那就只能传入 SRV_SMS_SIM_1;第四个参数(callback_func)是回调函数,短信发送完成后,会调用这个函数,我们可以在里面检测函数发送是否成功;第五个参数

(user_data)是发送短信携带的数据，这个数据并不会通过短信发送，而是会传入回调函数中，如果不需要携带数据，则传入 NULL。

以上两个函数接口一个用于发送 UCS2 字符编码的短信，一个用于发送 ASC 字符编码的短信，实际上通过短信发送的字符内容可以是任意编码格式，只不过非 UCS2 编码格式的字符无法在 MTK 设备的屏幕上直接显示。

接下来我们在代码中实现发送短信的功能，在 SCR_ID_HELLO_WORLD 屏幕中注册左软键事件用于发送短信，在 MyStudyAppMain.c 文件中修改代码如下(代码清单 5.1-1)：

---代码清单 5.1-1---

```
/*省略部分代码*/
#include "SmsSrvGprot.h"
/*省略部分代码*/
void mmi_send_sms_callback(srv_sms_callback_struct* callback_data)
{
    kal_prompt_trace(MOD_XDM, "--%d(%d)--%s--", __LINE__, callback_data->result, __FILE__);
}


void mmi_mystudy_send_sms(void)
{
    srv_sms_send_ucs2_text_msg(L"hello MTK!", L"10086", SRV_SMS_SIM_1,
                    mmi_send_sms_callback, NULL);
}


void mtk_helloworld_exit(void)
{
    /*省略部分代码*/
}
void mtk_helloworld(void)
{
    /*省略部分代码*/
    gui_BLT_double_buffer(0, 0, UI_DEVICE_WIDTH, UI_DEVICE_HEIGHT);

    SetLeftSoftkeyFunction(mmi_mystudy_send_sms, KEY_EVENT_UP);
    SetRightSoftkeyFunction(mmi_frm_scrn_close_active_id, KEY_EVENT_UP);
}
```

短信发送完后，可以在回调函数 mmi_send_sms_callback 中查看 callback_data->result 值，如果为 MMI_TRUE，则表示发送成功；如果为 MMI_FALSE，则表示发送失败。在模拟器上运行，如果短信发送失败，则需要在模拟器上打开 Catcher 工具，并点击"Start"按钮连上模拟器，如果有读者已经忘记了怎么操作，请翻阅前面讲解"下载和调试"的章节。如果购买了开发套件，可以下载到真机上运行，并把 mmi_mystudy_send_sms 函数放到

plutommi\mmi\MainMenu\MainMenuSrc\mainmenu.c 文件的 goto_main_menu 函数中运行，同时把代码中短信的接收号码“10086”改成自己能查阅短信的手机号码。这样可以简化操作流程，设备开机之后按下左软键(KEY_LSK)就会发送短信。在后续非界面相关的功能，如果放在真机上演示，都可以使用这种方式。

如今在物联网设备中，短信已经用的很少了，基本上都被网络代替了，因为短信传输的数据量比较少，而且相对于数据流量而言，运营商短信收费较贵；但是它也有一个好处，就是可以实现点对点传输，而且传输比较稳定。因此在有些产品中，依旧会通过短信来传输一些控制指令，此时我们就需要在系统接收短信的地方截取并解析。系统中只要有短信进来，就会执行 plutommi\mmi\Messages\MessagesSrc\SmsPsHandler.c 文件中的 mmi_sms_handle_new_msg_ind 函数，部分代码如图 5.1-1 所示，我们可以在这个函数中获取短信的内容以及发送短信的电话号码。

```
mmi_ret mmi_sms_handle_new_msg_ind(mmi_event_struct *evt)
{
    /*----------------------------------------------------------------*/
    /* Local Variables                                                */
    /*----------------------------------------------------------------*/
    srv_sms_event_struct* event_data = (srv_sms_event_struct*)evt;
    srv_sms_event_new_sms_struct* event_info;

    /*----------------------------------------------------------------*/
    /* Code Body                                                      */
    /*----------------------------------------------------------------*/
    event_info = (srv_sms_event_new_sms_struct*)event_data->event_info;

    if (event_info->msg_id == SRV_SMS_INVALID_MSG_ID)
    {
        mmi_msg_handle_class0_msg_ind(event_info);
    }
    else
    {
        mmi_msg_new_msg_ind(event_info);
        {
            /* send indication to UM */
            srv_um_refresh_ind_struct *refresh_ind;
            srv_sms_new_msg_struct *msg_data = (srv_sms_new_msg_struct*)event_info->msg_data;
```

图 5.1-1

这个函数中定义了两个变量 srv_sms_event_new_sms_struct* event_info 和 srv_sms_new_msg_struct *msg_data，如图 5.1-1 中红色箭头所示。短信内容保存在 event_info->content 中，编码格式为 UCS2；发送短信的电话号码保存在 msg_data->number 中，编码格式为 ASC。读者可以在模拟器中使用 Catcher 工具发送短信，通过断点查看其中的内容。

5.1.2 通话

在智能穿戴设备中通话功能修改最多的就是通话界面的定制。通话的整个过程分为去电(电话拨出但未接通)、来电(电话打进来但未接通)、接通、挂断，每个过程对应的界面 UI 都是不一样的。关于通话相关的界面定制，本节不作详细讲解，只是告诉大家相关界面的入口函数，以便大家自己修改时能够快速地找到地方。

通话相关的 UI 通常都在 plutommi\mmi\Ucm\UcmSrc\UcmUi.c 文件中。当电话打进来的时候会执行 mmi_ucm_enter_incoming_call 函数；当电话拨出去的时候会执行 mmi_ucm_enter_outgoing_call 函数；当电话接通时会执行 mmi_ucm_enter_in_call 函数；当电话挂断后会执行 mmi_ucm_release_ind 函数。

在实际开发中，除了界面定制，还需要拨打电话。拨打电话的功能其实挺简单的，就一个函数 void MakeCall(CHAR* ucs2_strNumber)，传入 UCS2 编码的电话号码就可以了。在 goto_main_menu 函数中添加一行代码 MakeCall((CHAR*)L"10086")，模拟器或真机设备开机后按下左软键就拨打 10086。代码如下(代码清单 5.1-2)：

---------------------------------------代码清单 5.1-2---------------------------------------

```
/*省略系统默认代码*/
#if defined(_MYSTUDY_APPLICATION_)
extern void mmi_mystudy_send_sms(void);
#endif
/*省略函数注释说明*/
void goto_main_menu(void)
{
    /*----------------------------------------------------------*/
    /* Local Variables                                          */
    /*----------------------------------------------------------*/


    /*----------------------------------------------------------*/
    /* Code Body                                                */
    /*----------------------------------------------------------*/
#if defined(_MYSTUDY_APPLICATION_)
//    mmi_mystudy_send_sms();
    MakeCall((CHAR*)L"10086");/*拨打电话*/
     #endif
/*省略系统默认代码*/
```

以上代码注释了发送短信的函数，SIM 卡通信是按优先级执行的，从高到底依次为：通话、短信、网络，并且这些通信事件不能同时进行，也就是说，当设备正在打电话时，无法发送短信也无法上网。另外需要注意的是 MakeCall 函数默认只会使用 SIM 卡 1 拨号，如果设备支持多个 SIM 卡，我们把卡插在了其他卡槽上，SIM 卡 1 的卡槽是空的，则调用这个函数无法拨打电话。如果要修改这个功能可以修改 MakeCall 函数中的 call_type 变量值。比如设备中插入了多张 SIM 卡，在拨号的时候提示用户选择 SIM 卡，可以把 call_type 的值改为 SRV_UCM_CALL_TYPE_NO_DATA_CSD。MakeCall 函数定义在 plutommi\mmi\Ucm\UcmSrc\UcmInterface.c 文件中，代码如图 5.1-2 所示。

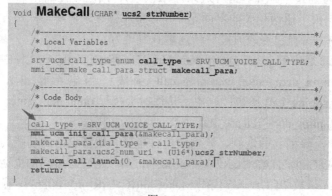

图 5.1-2

5.2 Socket 网络编程

人类社会已经进入了物联网时代，所谓"物联网"就是物物相连的互联网，其英文名称是"Internet of things"，简称"IoT"，这其中有两层意思：其一，物联网的核心和基础仍然是互联网，是在互联网基础上的延伸和扩展的网络；其二，其用户端延伸和扩展到了任意物品与物品之间进行信息交换和通信，也就是物物相息。物联网通过智能感知、识别技术等通信感知技术，广泛应用于网络的融合中，因此被称为继计算机、互联网之后世界信息产业发展的第三次浪潮。物联网是互联网的应用拓展，与其说物联网是网络，不如说物联网是业务和应用。

在我们的生活中，物联网设备已经无处不在，比如满大街的共享单车，可以说是最成功的物联网应用；其次各种智能手表、宠物防丢器、电瓶车防盗器等也都是很常见的物联网设备。对于个人用户而言，通常都是通过手机与物联网设备进行互联，比如智能手表，我们不仅可以实时查看手表上的状态信息(比如地理位置)，还可以实时下发操作指令，对手表进行控制。然而所有这些功能的实现，都是基于网络，只有在网络建立成功的情况下，才可以实现数据传输，这就是我们本章要讲的内容——Socket 网络编程。

MTK 设备上网与 PC 端上网有点区别，它使用的是"移动网络"，需 SIM 卡和 MTK 平台同时支持上网功能。要注意的是，这里的"移动网络"并非指中国移动通信集团公司提供的上网服务，而是指供所有可移动设备上网的服务。虽然联网方式不一样，但 Socket 网络编程方式是类似的，只不过 MTK 系统中对一些操作实现了封装，让我们使用起来会更简便一点。

Socket 网络连接功能可以当做一个 APP(application)来做，对它进行封装，以后在项目中只要用到网络发送数据的地方都可以调用它。在开始编码之前，先在 plutommi\MyStudyApp 目录下参考"MyStudyAppMain"创建一个 APP 文件夹，命名为"Socket"，里面包含两个子文件夹"Inc"、"Src"，两个子文件夹下分别再新建对应的文件"Socket.h"、"Socket.c"。目录结构如图 5.2-1 所示。(因用到的资源较少，就不单独创建资源目录了。)

图 5.2-1

然后在 plutommi\mmi\Inc\MMI_features.h 文件中定义 APP 功能宏，代码如代码清单 5.2-1 所示。

--代码清单 5.2-1--

```
/*省略系统代码*/

/*紧急联系人*/
#if defined(CFG_MMI_PHB_ICE_CONTACT) && ((CFG_MMI_PHB_ICE_CONTACT = _ON_)||
(CFG_MMI_PHB_ICE_CONTACT == __AUTO__))
    #ifndef __MMI_PHB_ICE_CONTACT__
    #define __MMI_PHB_ICE_CONTACT__
    #endif
#endif


#if defined(__MYSTUDY_APPLICATION__)/*自定义功能宏*/

/*socket 网络编程*/
#if defined(CFG_MMI_MYSTUDY_SOCKET) && (CFG_MMI_MYSTUDY_SOCKET!=_OFF_)
    #define __MMI_MYSTUDY_SOCKET__
#endif

#endif/*__MYSTUDY_APPLICATION__*/

#endif /* __MMI_FEATURES__ */
```

以上代码中 _MMI_MYSTUDY_SOCKET_ 是 Socket 的功能宏，而 CFG_MMI_MYSTUDY_SOCKET 是功能宏的开关。从代码中可以看出，如果宏开关为__OFF__，则宏 __MMI_MYSTUDY_SOCKET__ 是不会被定义的。在 MMI_features.h 中的宏定义，都是 MMI 层的功能宏，每个宏都有一个宏开关，而这些开关都定义在 plutommi\Customer\CustResource\HEXING60A_11B_MMI\MMI_features_switchHEXING60A_11B.h 文件中(注意：目录和文件名与项目名称相关)。为了区别系统宏开关和自己定义的宏开关，我们在 plutommi\mmi\Inc 目录下新建一个头文件"MMI_features_switch_mystudy_app.h"，里面全部存放自己定义的宏开关，代码如下(代码清单 5.2-2)：

--代码清单 5.2-2--

```
/*****************************************************************
 *    该文件用于放置自定义的功能宏开关
 *****************************************************************/
#ifndef __MMI_FEATURES_SWITCH_MYSTUDY_APP_H__
#define __MMI_FEATURES_SWITCH_MYSTUDY_APP_H__
```

```
#if defined(__MYSTUDY_APPLICATION__)

#include "MMI_features_type.h"

/*
    Description: socket 网络编程
    Option: [__ON__, __OFF__]
*/
#define CFG_MMI_MYSTUDY_SOCKET            (__ON__)

#endif /*__MYSTUDY_APPLICATION__*/
#endif /* __MMI_FEATURES_SWITCH_MYSTUDY_APP_H__ */
```

然后把这个头文件包含到 MMI_features_switchHEXING60A_11B.h 文件中。代码如下(代码清单 5.2-3)：

--代码清单 5.2-3--
```
#ifndef __MMI_FEATURES_SWITCH_H__
#define __MMI_FEATURES_SWITCH_H__

/***********************************************************************
* Option Value Definition
***********************************************************************/
#include "MMI_features_type.h"
#if defined(_MYSTUDY_APPLICATION_)
#include "MMI_features_switch_mystudy_app.h"
#endif
/***********************************************************************
* Switch Description
***********************************************************************/
/*省略系统代码*/
```

最后把新建的 Socket 相关的文件添加到 MyStudyApp 模块中，在 MyStudyApp.mak 文件中修改代码如下(代码清单 5.2-4)：

--代码清单 5.2-4--
```
/*省略部分代码*/
ifneq ($(filter __MYSTUDY_APPLICATION__ , $(strip $(MODULE_DEFS))), )
    INC_DIR += plutommi\MyStudyApp\MyStudyAppMain\Inc     #指定头文件路径
    SRC_LIST +=plutommi\MyStudyApp\MyStudyAppMain\Src\MyStudyAppMain.c#新增源文件
```

```
            ifneq ($(filter _MMI_MYSTUDY_SOCKET_, $(strip $(MODULE_DEFS))), )#添加 socket APP
                INC_DIR += plutommi\MyStudyApp\Socket\Inc
                SRC_LIST += plutommi\MyStudyApp\Socket\Src\Socket.c
            endif
        endif
```

依次执行 make new、make gen_modis 编译代码并重新生成模拟器，接下来就可以在 Socket app 相关的文件中编写 Socket 代码了。Socket 编程总共分为以下五个步骤，步骤中的示例代码不完整，在后面会把完整的代码全部贴出来。

(1) 获取 account id。

MTK 设备是基于 SIM 卡上网的，而在中国 SIM 卡目前有三家运营商：移动、电信、联通。在上网之前，我们要先确定上网的数据账户，而这个数据账户就是 account id。生成 account id 首先调用函数 cbm_register_app_id_with_app_info 注册一个 app id，然后调用 cbm_encode_data_account_id 函数返回 account id，我们直接把它封装成函数接口，代码如下（代码清单 5.2-5）：

---------------------------------------代码清单 5.2-5---------------------------------------

```c
U32 mmi_socket_get_account_id(void)
{
    cbm_app_info_struct app_info = {0x00};
    cbm_sim_id_enum sim = CBM_SIM_ID_TOTAL;

    if(0 == g_socket_data.account_id)
    {
        switch(g_socket_data.sim_id)
        {
            case MMI_SIM1:
            {
                sim = CBM_SIM_ID_SIM1;
                break;
            }
            case MMI_SIM2:
            {
                sim = CBM_SIM_ID_SIM2;
                break;
            }
            case MMI_SIM3:
            {
                sim = CBM_SIM_ID_SIM3;
                break;
            }
```

```
            case MMI_SIM4:
            {
                sim = CBM_SIM_ID_SIM4;
                break;
            }
            default:
            {
                return 0;
            }
        }

        app_info.app_icon_id = 0;
        app_info.app_str_id = 0;
        app_info.app_type = DTCNT_APPTYPE_BRW_HTTP|DTCNT_APPTYPE_
                            MRE_WAP|DTCNT_
APPTYPE_MRE_NET|DTCNT_APPTYPE_DEF;
        cbm_register_app_id_with_app_info(&app_info, (U8*)&g_socket_data.app_id);
            g_socket_data.account_id = cbm_encode_data_account_id(CBM_DEFAULT_ACCT_ID,
sim, g_socket_data.app_id, MMI_FALSE); /*获取 account id*/
        }

        return   g_socket_data.account_id;
    }
```
---

(2) 创建 socket id, 设置 Socket 属性。

Socket 相关的函数接口都包含在 interface\inet_ps\soc_api.h 文件中, 其函数体在 MTK 系统中都被封装起来了, 但在 MoDIS_VC9\drv_sim\src\w32_socket.c 文件中提供了一整套完整的同名函数接口。我们在模拟器上编写的 Socket 代码, 在真机上一般也可以运行。创建 Socket 的代码如下(代码清单 5.2-6):

---代码清单 5.2-6---

```
static kal_int8 mmi_socket_create(const U32 account_id)
{
    U8 socket_opt = 1;
    kal_int8 soc_id   = -1;

    soc_id = soc_create(SOC_PF_INET,SOC_SOCK_STREAM, 0, MOD_MMI, account_id);
                        /*创建 socket id*/
    if (soc_id < 0)
    {
        return SOC_INVAILD_ID;
```

```
        }
        socket_opt = KAL_TRUE;           /*设置异步属性*/
        if (soc_setsockopt(soc_id, SOC_NBIO, &socket_opt, sizeof(socket_opt)) < 0)
        {
            return SOC_SET_OPT_ERROR;
        }
        socket_opt = SOC_READ | SOC_WRITE | SOC_CONNECT | SOC_CLOSE; /*设置非阻塞属性*/
        if (soc_setsockopt(soc_id, SOC_ASYNC, &socket_opt, sizeof(socket_opt)) < 0)
        {
            return SOC_SET_OPT_ERROR;
        }
        return soc_id;
    }
```

创建 socket id 使用的函数接口为 kal_int8 soc_create(kal_uint8 domain, socket_type_enum type, kal_uint8 protocol, module_type mod_id, kal_uint32 nwk_account_id)。该函数的第一个参数(domain)是网络连接协议，目前只支持 Internet，只能传入 SOC_PF_INET；第二个参数(type)表示网络连接类型，通常使用的都是 TCP 连接，传入 SOC_SOCK_STREAM；第三个参数(protocol)目前没用，传入 0 就可以了；第四个参数(mod_id)为系统模块 id 名称，用于底层的消息发送和接收，我们的 APP 通常都是属于 MMI 层应用，所以传入 MOD_MMI；第五个参数(nwk_account_id)，就是上一步骤中获取的 account id。

socket id 创建成功后，紧接着把 Socket 属性设置为异步非阻塞。这里代码几乎是固定的，我们要分两次调用 soc_setsockopt 函数。

(3) 建立连接。

我们通常看到的服务器地址都类似于 "www.baidu.com" 这样的字符串，这个只是服务器的域名。Socket 与服务器的连接最终是通过 IP 地址建立起来的。如果服务器地址为 IP 地址，可以直接调用 soc_connect 建立连接；如果服务器地址为域名，则需要先调用 soc_gethostbyname 函数解析域名，并设置域名解析的消息(MSG_ID_APP_SOC_GET_HOST_BY_NAME_IND)处理函数，然后在处理函数中得到 IP 地址再建立连接。代码如下(代码清单 5.2-7):

--代码清单 5.2-7--
```
/*域名解析消息回调函数*/
static void mmi_socket_get_host_name_cb(void *msg)
{
    kal_int32 result = 0;
    app_soc_get_host_by_name_ind_struct* dns_ind = NULL;
    sockaddr_struct server_addr = {0x00};

    dns_ind = (app_soc_get_host_by_name_ind_struct *)msg;
```

```
            if (dns_ind->result == KAL_TRUE)
            {
                    memcpy((char *)&server_addr.addr, (char*)dns_ind->addr, dns_ind->addr_len);
                    server_addr.addr_len = dns_ind->addr_len;
            }

            result = soc_connect(dns_ind->request_id, &server_addr);
}

enum_soc_state mmi_socket_send_data(mmi_sim_enum sim_id, void *send_data, U32 data_bytes)
{
        /*省略部分代码*/
        /*判断服务器地址 server_addr 是否为 IP 地址*/
        if(soc_ip_check(server_addr, soc_addr.addr, &ip_validity)=KAL_FALSE || KAL_FALSE==ip_validity)
        {
                /*如果不是 IP 地址，则先解析域名*/
                resul =soc_gethostbyname(KAL_FALSE, MOD_MMI, g_socket_data.soc_id, server_addr,
(kal_uint8*) soc_addr.addr, (kal_uint8 *)&soc_addr.addr_len, 0, g_socket_data.account_id);
                if (SOC_SUCCESS != result)
                {
                        if (result == SOC_WOULDBLOCK)
                        {
                                /*设置域名解析的消息回调函数*/

            mmi_frm_set_protocol_event_handler(MSG_ID_APP_SOC_GET_HOST_BY_NAME_IND,
(PsIntFuncPtr)mmi_socket_get_host_name_cb, TRUE);
                                return SOC_CREATE_SUCCESS;
                        }
                }
        }
        /*如果服务器地址为 IP 地址，或者解析域名返回 SOC_SUCCESS，则直接建立连接*/
        result = soc_connect(g_socket_data.soc_id, &soc_addr);
return SOC_CONNECT_FAILED;
}
```

----

kal_int8 soc_connect(kal_int8 s, sockaddr_struct *addr)函数的第一个参数(s)为 socket id，第二个参数(addr)为服务器地址的结构体 sockaddr_struct 变量。sockaddr_struct 结构体定义如图 5.2-2 所示。

```
/* socket address structure */
typedef struct
{
    socket_type_enum    sock_type;  /* socket type */
    kal_int16           addr_len;   /* address length */
    kal_uint16          port;       /* port number */
    kal_uint8           addr[MAX_SOCK_ADDR_LEN];
    /* IP address. For keep the 4-type boundary,
     * please do not declare other variables above "addr"
     */
} sockaddr_struct;
```

图 5.2-2

其中 sock_type 表示联网类型,必须与 soc_create 函数的第二个参数(type)值一致;addr_len 表示 IP 地址的长度,目前使用的都是 IPV4 地址,所以长度为 4 字节;port 是服务的端口号;addr 是服务器的 IP 地址,需注意的是,这里的 IP 地址并不是以字符串形式存储的,而是以字节数值存储的,比如 "192.168.1.1",存储为 addr[0]=192,addr[1]=168,addr[2]=1,addr[3]=1。

kal_int8 soc_gethostbyname(kal_bool is_blocking, module_type mod_id, kal_int32 request_id, const kal_char *domain_name, kal_uint8 *addr, kal_uint8 *addr_len, kal_uint8 access_id, kal_uint32 nwk_account_id)第一个参数(is_blocking)表示是否为阻塞式调用,此处只能传入 FALSE,因为 MTK 平台不支持阻塞式调用;第二个参数(mod_id)为接收消息的模块 id,同 soc_create 函数中的参数 mod_id;第三个参数(request_id)是我们自定义的 id 值,用于区分是哪个请求,标识不同的应用,它与域名解析的消息处理函数中 dns_ind->request_id 值一致,如果根据 socket id 来区分请求,则可以直接传入 socket id;第四个参数(domain_name)为服务器的域名,比如 "www.baidu.com";第五个参数(addr)存储解析域名得到的 IP 地址;第六个参数(addr_len)表示解析域名后得到的 IP 地址字节长度,只有函数返回 SOC_SUCCESS 时第五、六个参数才会被赋值;第七个参数(access_id)是访问控制 id,我们没有就直接传入 0;第八个参数(nwk_account_id)是前面获取到的 account id。

(4) 收发数据。

在连接服务器或解析域名的时候,一般都会返回 SOC_WOULDBLOCK 状态,这个状态表示函数正在执行,但不能立即获得执行结果,需要等待。此时我们就需要设置对应的消息处理函数,来接收函数执行的结果,就像上一步解析域名时,需设置 MSG_ID_APP_SOC_GET_HOST_BY_NAME_IND 消息的处理函数来处理解析域名的结果一样。在建立连接的时候,也要设置 MSG_ID_APP_SOC_NOTIFY_IND 消息的处理函数,如果连接成功,系统就会调用这个函数,然后我们就可以在这个函数里面监听消息类型,实现收发数据。比如设置 MSG_ID_APP_SOC_NOTIFY_IND 消息的处理函数为 socket_soc_notify_handle,则语句如下:

    mmi_frm_set_protocol_event_handler(MSG_ID_APP_SOC_NOTIFY_IND, socket_soc_notify_handle, MMI_TRUE);

在 socket_soc_notify_handle 函数中解析 Socket 的不同事件 id,这些事件一般为 SOC_READ(接收数据)、SOC_CONNECT(连接服务器)、SOC_WRITE(发送数据)、SOC_CLOSE(关闭 Socket)。我们接收数据的操作就在收到 SOC_READ 时进行,调用接口

soc_recv，发送数据的操作就在收到 SOC_CONNECT 或 SOC_WRITE 时进行，调用接口 soc_send。socket_soc_notify_handle 函数的代码如下(代码清单 5.2-8)：

---代码清单 5.2-8---

```
static void mmi_socket_soc_notify_handle(void* msg)
{
    app_soc_notify_ind_struct *soc_notify = (app_soc_notify_ind_struct*)msg;
    kal_int8 result = 0;
    switch (soc_notify->event_type)
    {
        case SOC_READ:/*接收数据*/
        {
            soc_recv(socket_id, recv_data_buff, recv_buffer_size, 0);
            break;
        }
        case SOC_CONNECT:/*连接成功*/
        case SOC_WRITE: /*发送数据*/
        {
            soc_send(socket_id, send_data_buff, send_data_size, 0);
            break;
        }
        case SOC_CLOSE: /*关闭 socket*/
        {
            soc_close(socket_id);
            socket_id =  -1;
            break;
        }
    }
}
```

发送数据使用的函数接口为 kal_int32 soc_send(kal_int8 s, void *buf, kal_int32 len, kal_uint8 flags)，函数的第一个参数(s)为 socket id；第二个参数(buf)为要发送的数据；第三个参数(len)为要发送数据的字节个数；第四个参数(flag)我们传入 0 就可以了。接收数据的函数接口为 kal_int32 soc_recv(kal_int8 s, void *buf, kal_int32 len, kal_uint8 flags)，这个函数的参数与 soc_send 类似，第一个为 socket id；第二个为保存接收数据的 buffer，第三个为 buffer 的字节大小，第四个也是传入 0。

(5) 关闭 Socket。
如果设备的网络信号不好或者网络连接超时，又或者服务器主动断开连接，MSG_ID_APP_SOC_NOTIFY_IND 消息处理函数中都会收到 SOC_CLOSE 消息，如代码清单 5.2-8 所以，此时就要调用 kal_int8 soc_close(kal_int8 s)函数关闭 Socket 释放资源，这个函数只需要传入socket id，需特别注意的是，Socket 关闭之后，要把保存 socket id 的变量置为–1，因为 socket id 的有效取值范围为 0～127，置为–1 有助于我们对其有效性的判断。

除了 SOC_CLOSE 消息之外，我们在 Socket 运行的过程中碰到的一些错误，也应该关闭 Socket，以便下次重连。关于 Socket 的错误信息，可以到 interface\inet_ps 目录下的一些头文件中去查找，比如 soc_consts.h 文件。

在 MTK 平台上，同一时间通常只支持一个 Socket 连接，如果要同时支持多个连接，需要 MTK 公司释放补丁。但是一个 Socket 连接对于我们来说也基本上够用了，可以把所有数据都放到队列里面依次发送。另外由于移动设备的网络比较慢，特别是 2G 网络，发送数据的频率也不能太高，笔者曾经尝试过 1 秒发送一次，仍然很容易造成网络阻塞而导致网络断开或者造成假连接，任何数据都发不出去，严重的甚至导致死机，因此建议数据传输的频率最低为 2 秒。

以下附上 Socket.h 和 Socket.c 文件的完整源码，这个 Socket 代码是笔者封装过的，对外的函数接口只有三个，一个初始化函数，一个发送数据的接口，一个设置回调函数的接口。这样的封装可以让我们的代码高度模块化，实现"高内聚，低耦合"的软件设计标准，分别如代码清单 5.2-9 和代码清单 5.2-10 所示。

```
------------------------------------------Socket.h 代码清单 5.2-9------------------------------------------
#ifndef __SOCKET_H__
#define __SOCKET_H__


#include "MMI_features.h"


#if defined(__MMI_MYSTUDY_SOCKET__)


/*************************************************************
* 头文件
*************************************************************/
#include "soc_api.h"
#include "MMIDataType.h"
#include "gui.h"
#include "nvram_user_defs.h"
#include "ssl_api.h"


/*************************************************************
* 宏定义
*************************************************************/
#define NETWORK_DEFAULT_SERVER_ADDR          "172.20.65.107"      /*服务器地址*/
#define NETWORK_DEFAULT_SERVER_PORT          8080                 /*端口号*/


#define HTTP_GET_DATA_FORMAT "GET %s%s HTTP/1.1\r\nHost:%s:%d\r\nAccept: */*\r\nConnection:close\r\nUser-Agent:%s\r\n\r\n"
```

```c
#define HTTP_POST_DATA_FORMAT "POST %s HTTP/1.1\r\nHost:%s:%d\r\nAccept:*/*\r\nConnection:close\r\nUser-Agent:%s\r\nContent-Length:%d\r\n\r\n%s"

#define SOC_SEND_BUFFER_SIZE          (512)/*保存发送数据的内存大小*/
#define SOC_RECV_BUFFER_SIZE          (512)/*保存接收数据的内存大小*/

/***************************************************************
* 数据类型
***************************************************************/

/*socket 执行的状态*/
typedef enum{
    SOC_SUCCESS_STATE_START = 0,         /*socket 没有出错的状态 id*/
    SOC_CREATE_SUCCESS,                  /*socket 创建成功*/
    SOC_CONNECT_SUCCESS,                 /*socket 连接成功*/
    SOC_SEND_SUCCESS,                    /*socket 发送成功*/
    SOC_RECV_SUCCESS,                    /*socket 接收成功*/
    SOC_CONNECT_RELEASE,                 /*socket 连接释放*/

    SOC_ERROR_STATE_START = SOC_SUCCESS_STATE_START,
    SOC_COMMON_ERROR = -1,               /*其他错误*/
    SOC_INVAILD_ID = -2,                 /*无效的 socket id*/
    SOC_SET_OPT_ERROR = -3,              /*设置 socket 属性失败*/
    SOC_GET_HOST_ERROR = -4,             /*域名解析错误*/
    SOC_CONNECT_FAILED = -5,             /*服务器连接错误*/
}enum_soc_state;

typedef void (*soc_net_cb)(enum_soc_state state, void *soc_data);

/*socket 数据*/
typedef struct __stu_network_soc{
    kal_int8 soc_id;                     /*socket id，有效值 >=0*/
    mmi_sim_enum sim_id;                 /*默认上网的 SIM 卡*/
    soc_net_cb callback;                 /*网络连接的回调函数*/
    MMI_BOOL online;                     /*网络是否在线*/
    U32 account_id;                      /* 0xFFFFFFFF*/
    U32 app_id;

    U8 send_data_buff[SOC_SEND_BUFFER_SIZE]; /*保存需要发送的数据*/
    U32 send_data_size;                  /*已经发送的数据的字节数*/
```

```c
    U8 recv_data_buff[SOC_RECV_BUFFER_SIZE];    /*保存接收到的数据*/
    U32 recv_data_size;                          /*已经接收的数据的字节数*/
}stu_socket_data;

/****************************************************************
* 函数声明
****************************************************************/

/****************************************************************
*功  能：设置网络连接的回调函数
*参  数：soc_cb [in] -- 网络连接的回调函数
*返回值：void
****************************************************************/
extern void mmi_socket_set_callback(soc_net_cb soc_cb);

/****************************************************************
* 功  能：连接网络，与服务器进行数据交互
* 参  数：
*         sim_id     [in] -- 联网的 sim 卡
*         send_data [in] -- 需要发送数据的内容
*         data_bytes[in] -- 发送数据的字节个数
* 返回值：
*         请查看 enum_soc_state 枚举
****************************************************************/
extern enum_soc_state mmi_socket_send_data(mmi_sim_enum sim_id, void *send_data, U32 data_bytes);

/****************************************************************
*功  能：初始化 g_socket_data 数组
*参  数：void
*返回值：void
****************************************************************/
extern void socket_socket_init(void);


#endif


#endif /* __SOCKET_H__ */
```

--Socket.c 代码清单 5.2-10--

```c
#include "mmi_features.h"

#if defined(__MMI_MYSTUDY_SOCKET__)

/*****************************************************************
* 头文件
*****************************************************************/
#include "Socket.h"
#include "cbm_consts.h"
#include "app2soc_struct.h"
#include "soc_api.h"
#include "DtcntSrvGprot.h"
#include "mmi_frm_events_gprot.h"
#include "mmi_frm_mem_gprot.h"
#include "mmi_frm_nvram_gprot.h"
#include "cbm_api.h"

/*****************************************************************
/* 宏常量定义
*****************************************************************/

/*****************************************************************
/* 全局变量
*****************************************************************/
static stu_socket_data g_socket_data={0x00};/*可以使用的 socket 数据*/

/*****************************************************************
/* 函数声明
*****************************************************************/
static void mmi_socket_release(void);
static void mmi_socket_get_soc_addr_msg(sockaddr_struct *soc_addr, kal_char *server_addr);
static void mmi_socket_soc_notify_handle(void* msg);

/*****************************************************************
/* 内部静态函数
*****************************************************************/

MMI_BOOL mmi_socket_get_online(void)
```

```c
{
    return g_socket_data.online;
}

void mmi_socket_set_online(MMI_BOOL status)
{
    g_socket_data.online = status;
}

/***************************************************************
*功    能：socket 发送消息
*参    数：void
*返回值：void
***************************************************************/
static S32 mmi_socket_soc_send_data(void)
{
    /*----------------------------------------------------------------*/
    /* Local Variables                                              */
    /*----------------------------------------------------------------*/
    S32 send_bytes=0;

    /*----------------------------------------------------------------*/
    /* Code Body                                                    */
    /*----------------------------------------------------------------*/
    if(g_socket_data.soc_id<0 || 0==g_socket_data.send_data_size)
    {
        return 0;
    }

    send_bytes = soc_send(g_socket_data.soc_id, g_socket_data.send_data_buff,
                    g_socket_data.send_data_size, 0);
    if(send_bytes > 0)
    {
        if(NULL != g_socket_data.callback)          /*执行回调函数，发送成功 通知*/
        {
            g_socket_data.callback(SOC_SEND_SUCCESS, &g_socket_data);
        }
    }
    else if(SOC_WOULDBLOCK==send_bytes)
    {
```

```c
            mmi_frm_set_protocol_event_handler(MSG_ID_APP_SOC_NOTIFY_IND,
                    mmi_socket_soc_notify_handle, MMI_TRUE);
    }
    else
    {
        if(NULL != g_socket_data.callback)
        {
            g_socket_data.callback(SOC_COMMON_ERROR, &g_socket_data);
        }
    }
    return send_bytes;
}


/*************************************************************************
*功  能：socket 接收消息
*参  数：void
*返回值：void
*************************************************************************/
static S32 mmi_socket_soc_recv_data(void)
{
    /*----------------------------------------------------------------*/
    /* Local Variables                                                */
    /*----------------------------------------------------------------*/
    S32 recv_bytes = -1;
    U8 *recv_buff_ptr = NULL, *temp_buff=NULL, *p1=NULL, *p2=NULL;

    /*----------------------------------------------------------------*/
    /* Code Body                                                      */
    /*----------------------------------------------------------------*/

    memset(g_socket_data.recv_data_buff, 0x00, SOC_RECV_BUFFER_SIZE);
    g_socket_data.recv_data_size = 0;
    recv_bytes = soc_recv(g_socket_data.soc_id, g_socket_data.recv_data_buff, SOC_RECV_
                    BUFFER_ SIZE, 0);
    if(recv_bytes > 0)
    {
        g_socket_data.recv_data_size = recv_bytes;
        if(NULL != g_socket_data.callback)
        {
            g_socket_data.callback(SOC_RECV_SUCCESS, &g_socket_data);
```

```
                    }
                }
                else
                {
                    if(SOC_WOULDBLOCK==recv_bytes)/*-2*/
                    {
                        mmi_frm_set_protocol_event_handler(MSG_ID_APP_SOC_NOTIFY_IND,
                            mmi_socket_soc_notify_handle, MMI_TRUE);
                    }
                    else
                    {
                        if(NULL != g_socket_data.callback)
                        {
                            g_socket_data.callback(SOC_COMMON_ERROR, &g_socket_data);
                        }
                    }
                }

                return recv_bytes;
}

/***********************************************************************
*功    能：socket 信号的回调函数
*参    数：msg [in] -- 信号对应的信号体指针
*返回值：
*          void
***********************************************************************/
static void mmi_socket_soc_notify_handle(void* msg)
{
    /*----------------------------------------------------------------*/
    /* Local Variables                                                */
    /*----------------------------------------------------------------*/
    app_soc_notify_ind_struct *soc_notify = (app_soc_notify_ind_struct*)msg;
    kal_int8 result = 0;

    /*----------------------------------------------------------------*/
    /* Code Body                                                      */
    /*----------------------------------------------------------------*/
    if(NULL == soc_notify)
    {
```

```
        return;
    }

    if(SOC_BEARER_FAIL == soc_notify->error_cause)
    {
        mmi_socket_release();
        return;
    }

    mmi_socket_set_online(MMI_TRUE);/*有数据发送的时候,标志 GPRS 为激活状态*/

    /*第 4 步: 收发数据*/
    switch (soc_notify->event_type)
    {
        case SOC_READ:
        {
            mmi_socket_soc_recv_data();        /*接收数据*/
            break;
        }
        case SOC_CONNECT:
        case SOC_WRITE:
        {
            mmi_socket_soc_send_data();        /*发送数据*/
            break;
        }
        case SOC_CLOSE:
        {
            mmi_socket_release();
            break;
        }
        default:
        {
            break;
        }
    }
}

/******************************************************************
*功  能:socket 通过 dns 得到当前服务器的 ip 地址的回调函数
*参  数:msg [in] -- 信号对应的信号体指针
```

*返回值：
*   void
**/
static void mmi_socket_get_host_name_cb(void *msg)
{
    /*--*/
    /* Local Variables   */
    /*--*/
     kal_int32 result = 0;
     app_soc_get_host_by_name_ind_struct* dns_ind = NULL;
    sockaddr_struct server_addr = {0x00};

    /*--*/
    /* Code Body   */
    /*--*/
     if (msg == NULL)
     {
          return;
     }

     dns_ind = (app_soc_get_host_by_name_ind_struct *)msg;
      mmi_frm_clear_protocol_event_handler(MSG_ID_APP_SOC_GET_HOST_BY_NAME_IND, (PsIntFuncPtr)mmi_socket_get_host_name_cb);

     mmi_socket_get_soc_addr_msg(&server_addr, NULL);
     if (dns_ind->result == KAL_TRUE)
     {
           memcpy((char *)&server_addr.addr, (char*)dns_ind->addr, dns_ind->addr_len);
           server_addr.addr_len = dns_ind->addr_len;
     }

     result = soc_connect(dns_ind->request_id, &server_addr);
    if(result==SOC_SUCCESS)/*直接发送数据*/
     {
         mmi_socket_soc_send_data();
         return;
     }
     else if (result == SOC_WOULDBLOCK)
     {

```
            mmi_frm_set_protocol_event_handler(MSG_ID_APP_SOC_NOTIFY_IND,
                                   mmi_socket_soc_notify_handle, MMI_TRUE);
            return;
        }
        else/*出错了*/
        {
            if(NULL != g_socket_data.callback)
            {
                g_socket_data.callback(SOC_CONNECT_FAILED, &g_socket_data);
            }
            mmi_socket_release();
            return;
        }
}


/***************************************************************
*功  能：获取 account id
*参  数：void
*返回值：
*         account_id
****************************************************************/
U32 mmi_socket_get_account_id(void)
{
    /*-----------------------------------------------------------*/
    /* Local Variables                                           */
    /*-----------------------------------------------------------*/
    cbm_app_info_struct app_info = {0x00};
    cbm_sim_id_enum sim = CBM_SIM_ID_TOTAL;


    /*-----------------------------------------------------------*/
    /* Code Body                                                 */
    /*-----------------------------------------------------------*/
    if(0 == g_socket_data.account_id)
    {
        switch(g_socket_data.sim_id)
        {
            case MMI_SIM1:
            {
                sim = CBM_SIM_ID_SIM1;
                break;
```

```c
            }
            case MMI_SIM2:
            {
                sim = CBM_SIM_ID_SIM2;
                break;
            }
            case MMI_SIM3:
            {
                sim = CBM_SIM_ID_SIM3;
                break;
            }
            case MMI_SIM4:
            {
                sim = CBM_SIM_ID_SIM4;
                break;
            }
            default:
            {
                return 0;
            }
        }

        app_info.app_icon_id = 0;
        app_info.app_str_id = 0;
        app_info.app_type = DTCNT_APPTYPE_BRW_HTTP|DTCNT_APPTYPE_MRE_
                    WAP|DTCNT_APPTYPE_MRE_NET|DTCNT_APPTYPE_DEF;
        cbm_register_app_id_with_app_info(&app_info, (U8*)&g_socket_data.app_id);
        g_socket_data.account_id  =  cbm_encode_data_account_id(CBM_DEFAULT_ACCT_ID, sim, g_socket_data.app_id, MMI_FALSE);
    }

    return  g_socket_data.account_id;
}


/******************************************************************
*功    能：创建 socket
*参    数：account_id [in] -- 数据账户
*返回值：
*         socket id
******************************************************************/
```

```c
static kal_int8 mmi_socket_create(const U32 account_id)
{
    /*----------------------------------------------------------*/
    /* Local Variables                                          */
    /*----------------------------------------------------------*/
    U8 socket_opt = 1;
    kal_int8 soc_id   = -1;

    /*----------------------------------------------------------*/
    /* Code Body                                                */
    /*----------------------------------------------------------*/
    soc_id = soc_create(SOC_PF_INET, SOC_SOCK_STREAM, 0, MOD_MMI, account_id);
    if (soc_id < 0)
    {
        return SOC_INVAILD_ID;
    }

    socket_opt = KAL_TRUE;
    if (soc_setsockopt(soc_id, SOC_NBIO, &socket_opt, sizeof(socket_opt)) < 0)
    {
        return SOC_SET_OPT_ERROR;
    }
    socket_opt = SOC_READ | SOC_WRITE | SOC_CONNECT | SOC_CLOSE;
    if (soc_setsockopt(soc_id, SOC_ASYNC, &socket_opt, sizeof(socket_opt)) < 0)
    {
        return SOC_SET_OPT_ERROR;
    }

    return soc_id;
}

/***********************************************************************
*功  能：释放 socket
*参  数：void
*返回值：void
***********************************************************************/
static void mmi_socket_release(void)
{
    /*----------------------------------------------------------*/
    /* Local Variables                                          */
```

```
/*-----------------------------------------------------------*/


/*-----------------------------------------------------------*/
/* Code Body                                                 */
/*-----------------------------------------------------------*/
mmi_socket_set_online(MMI_FALSE);
if(NULL != g_socket_data.callback)
{
    g_socket_data.callback(SOC_CONNECT_RELEASE, &g_socket_data);
}

if(g_socket_data.soc_id >= 0)
{
    soc_close(g_socket_data.soc_id);
    g_socket_data.soc_id = -1;
}

memset(g_socket_data.send_data_buff, 0x00, SOC_SEND_BUFFER_SIZE);
g_socket_data.send_data_size = 0;

memset(g_socket_data.recv_data_buff, 0x00, SOC_RECV_BUFFER_SIZE);
g_socket_data.recv_data_size = 0;
}


/*****************************************************************************
*功　能：获取服务器 信息
*参　数：soc_addr       [out]-- 服务器地址信息
*        server_addr [out]-- 服务器域名
*返回值：
*        MMI_FALSE / MMI_TRUE
*****************************************************************************/
static void mmi_socket_get_soc_addr_msg(sockaddr_struct *soc_addr, kal_char *server_addr)
{
    /*-----------------------------------------------------------*/
    /* Local Variables                                           */
    /*-----------------------------------------------------------*/
    S16 error = 0;


    /*-----------------------------------------------------------*/
    /* Code Body                                                 */
```

```c
    /*-----------------------------------------------------------*/
    if(NULL == soc_addr && NULL==server_addr)
    {
        return;
    }

    if(NULL != soc_addr)
    {
        soc_addr->sock_type = SOC_SOCK_STREAM;
        soc_addr->port = NETWORK_DEFAULT_SERVER_PORT;
    }
    if(NULL != server_addr)
    {
        strcpy((char*)server_addr, (char *)NETWORK_DEFAULT_SERVER_ADDR);
    }
    soc_addr->addr_len = 0x04;//soc_htonl()

    return;

}


/***********************************************************************
*功  能：设置 socket 发送数据
*参  数：
*        send_data [in] -- 需要发送数据的内容
*        data_bytes[in] -- 发送数据的字节个数
*返回值：
*        MMI_FALSE / MMI_TRUE
***********************************************************************/
static MMI_BOOL mmi_socket_set_send_content(void *send_data, U32 data_bytes)
{
    /*-----------------------------------------------------------*/
    /* Local Variables                                           */
    /*-----------------------------------------------------------*/


    /*-----------------------------------------------------------*/
    /* Code Body                                                 */
    /*-----------------------------------------------------------*/
    if(NULL==send_data || 0==data_bytes || data_bytes>SOC_SEND_BUFFER_SIZE)
    {
```

```c
        return MMI_FALSE;
    }
    memset(g_socket_data.send_data_buff, 0x00, SOC_SEND_BUFFER_SIZE);
    memcpy(g_socket_data.send_data_buff, send_data, data_bytes);
    g_socket_data.send_data_size = data_bytes;

    return MMI_TRUE;
}

/***********************************************************************
*功    能：设置网络连接的回调函数
*参    数：soc_cb [in] -- 网络连接的回调函数
*返回值：void
************************************************************************/
void mmi_socket_set_callback(soc_net_cb soc_cb)
{
    g_socket_data.callback = soc_cb;
}


/***********************************************************************
* 功    能：连接网络，与服务器进行数据交互
* 参    数：
*        sim_id      [in] -- 联网的 sim 卡
*        send_data [in] -- 需要发送数据的内容
*        data_bytes[in] -- 发送数据的字节个数
* 返回值：
*        请查看 enum_soc_state 枚举
************************************************************************/
enum_soc_state mmi_socket_send_data(mmi_sim_enum sim_id, void *send_data, U32 data_bytes)
{
    /*----------------------------------------------------------------*/
    /* Local Variables                                                */
    /*----------------------------------------------------------------*/
    sockaddr_struct soc_addr = {0x00};
    kal_bool ip_validity = KAL_FALSE;
    S32 result = 0;
    kal_char server_addr[128]={0x00};


    /*----------------------------------------------------------------*/
    /* Code Body                                                      */
```

```c
/*--------------------------------------------------------------*/
if(NULL==send_data || 0==data_bytes)
{
    return SOC_COMMON_ERROR;
}

g_socket_data.sim_id = sim_id;

if(g_socket_data.soc_id >= 0)        /*soc id 没有被释放, 直接发送数据*/
{
    if(MMI_FALSE == mmi_socket_set_send_content(send_data, data_bytes))
                    /*准备要发送的数据*/
    {
        return SOC_CONNECT_SUCCESS;
    }
    result = mmi_socket_soc_send_data();
    if(result < 0)
    {
        mmi_socket_release();
    }
    else
    {
        return SOC_CONNECT_SUCCESS;
    }
}

mmi_socket_get_soc_addr_msg(&soc_addr, server_addr);

g_socket_data.account_id = mmi_socket_get_account_id();   /*第 1 步: 获取 account id*/
g_socket_data.soc_id=mmi_socket_create(g_socket_data.account_id ); /*第 2 步: 创建 socket id*/
if(g_socket_data.soc_id < 0)
{
    if(NULL != g_socket_data.callback)
    {
        g_socket_data.callback(SOC_COMMON_ERROR, &g_socket_data);
    }
    mmi_socket_release();
    return SOC_INVAILD_ID;
}

if(MMI_FALSE == mmi_socket_set_send_content(send_data, data_bytes))/*准备要发送的数据*/
{
```

```c
        return SOC_CONNECT_SUCCESS;
    }

    /*第3部: 建立连接*/
    if(soc_ip_check(server_addr, soc_addr.addr, &ip_validity)==KAL_FALSE ||
            KAL_FALSE==ip_validity)
    {
        result = soc_gethostbyname(KAL_FALSE, MOD_MMI, g_socket_data.soc_id, server_addr,
(kal_uint8 *) soc_addr.addr, (kal_uint8 *)&soc_addr.addr_len, 0, g_socket_data.account_id);
        if (SOC_SUCCESS != result)
        {
            if (result == SOC_WOULDBLOCK)
            {
                mmi_frm_set_protocol_event_handler(MSG_ID_APP_SOC_GET_HOST_BY_
                    NAME_IND, (PsIntFuncPtr)mmi_socket_get_host_name_cb, TRUE);
                return SOC_CREATE_SUCCESS;
            }
            else/*连接出错, 释放 socket*/
            {
                if(NULL != g_socket_data.callback)
                {
                    g_socket_data.callback(SOC_COMMON_ERROR, &g_socket_data);
                }
                mmi_socket_release();
                return SOC_GET_HOST_ERROR;
            }
        }
    }

    result = soc_connect(g_socket_data.soc_id, &soc_addr);/*连接网络*/
    if(result==SOC_SUCCESS)/*连接成功或已经连接, 就直接发送数据*/
    {
        mmi_socket_soc_send_data();
        return SOC_CONNECT_SUCCESS;
    }
    else if (SOC_WOULDBLOCK == result)
    {
        mmi_frm_set_protocol_event_handler(MSG_ID_APP_SOC_NOTIFY_IND,
            mmi_socket_soc_notify_handle, MMI_TRUE);
        return SOC_CONNECT_SUCCESS;
    }
    else
```

```c
    {
        if(NULL != g_socket_data.callback)/*连接出错, 释放 socket*/
        {
            g_socket_data.callback(SOC_COMMON_ERROR, &g_socket_data);
        }
        mmi_socket_release();
    }

    return SOC_CONNECT_FAILED;
}

/****************************************************************
*功  能：初始化 g_socket_data 数组
*参  数：void
*返回值：void
****************************************************************/
void socket_socket_init(void)
{
    /*------------------------------------------------------------*/
    /* Local Variables                                          */
    /*------------------------------------------------------------*/
    /*------------------------------------------------------------*/
    /* Code Body                                                */
    /*------------------------------------------------------------*/
    memset(&g_socket_data, 0x00, sizeof(g_socket_data));
    g_socket_data.soc_id = -1;
    g_socket_data.sim_id = MMI_SIM1;
}

#endif
```

接下来我们运行以上代码，利用前面创建的 Reminder 定时器，一分钟发送一次数据。在 MyStudyAppMain.c 文件中修改代码如下(代码清单 5.2-11):

---代码清单 5.2-11---

```c
#include "Socket.h"
/*省略部分代码*/
void mmi_mystudy_socket_cb(enum soc_state state, stu_socket_data *soc_data)
{
    if(NULL == soc_data)
    {
        return;
```

```c
        }
        switch(state)
        {
            case SOC_SEND_SUCCESS:/*socket 发送成功*/
            {
                kal_prompt_trace(MOD_XDM, "--%d(%s)--%s--", _LINE_, soc_data->send_data_buff,
                        _FILE_);
                break;
            }
            case SOC_RECV_SUCCESS:/*socket 接收成功*/
            {
                kal_prompt_trace(MOD_XDM, "--%d(%s)--%s--", _LINE_, soc_data->recv_data_buff,
                        _FILE_);
                break;
            }
        }
}

mmi_ret mmi_mystudy_reminder_proc(mmi_event_struct *evt)
{
    srv_reminder_evt_struct *reminder_evt = (srv_reminder_evt_struct*)evt;
    MYTIME curr_time={0x00};
    U8 soc_data[64] = {0x00};

    /*省略部分代码*/
    switch (reminder_evt->notify)
    {
        case SRV_REMINDER_NOTIFY_INIT:    /*初始化*/
        {
            socket_socket_init();  /*socket 初始化，通常也可以放在 mmi_bootup_notify_
                    completed 函数中调用*/
            break;
        }
        case SRV_REMINDER_NOTIFY_EXPIRY:    /*触发*/
        {
            GetDateTime(&curr_time);
            sprintf(soc_data, "mystudy socket %04d/%02d/%02d %02d:%02d:%02d", curr_time.nYear,
        curr_time.nMonth, curr_time.nDay, curr_time.nHour, curr_time.nMin, curr_time.nSec);
            mmi_socket_set_callback(mmi_mystudy_socket_cb);/*设置 socket 回调函数*/
```

```
                mmi_socket_send_data(MMI_SIM1, soc_data, strlen(soc_data));
                mmi_set_reminder_time();
                break;
            }
            /*省略部分代码*/
        }
        return MMI_RET_OK;
    }
```

Socket 初始化函数 socket_socket_init 在开机的时候就应该执行，我们可以把它放在 mmi_mystudy_reminder_proc 函数的 SRV_REMINDER_NOTIFY_INIT 消息中调用，也可以放在 plutommi\mmi\Bootup\BootupSrc\BootupInitApps.c 文件的 mmi_bootup_notify_completed 函数中调用，这个函数每次系统开机完成之后都会执行。

在定时器触发的消息 SRV_REMINDER_NOTIFY_EXPIRY 中调用 mmi_socket_set_callback 函数设置 Socket 的回调函数，并调用 mmi_socket_send_data 函数发送数据。mmi_socket_send_data 函数中封装了整个 Socket 的实现方法，我们只需要传入联网 SIM 卡待发送的数据及数据长度就可以了，其他的一切操作都在函数内部完成。另外设置回调函数的目的，是为了把一些状态信息引到外部来处理，比如发送成功、接收成功、Socket 错误信息等，这样可以使 Socket 联网功能在不同的应用之间实现差异化处理，且不会影响已封装的 Socket 代码框架结构。

Socket 的代码已经写好了，但是服务器怎么办呢？这里给大家推荐一个工具"TCP/UDP Socket 调试工具"，打开后界面如图 5.2-3 所示，这个工具可以模拟一个本地服务器，方便我们调试代码，读者可以自己到网上下载。但如果在真机上调试网络，这个工具就不管用了，必须使用真实服务器才行。

图 5.2-3

使用 TCP/UCP Socket 调试工具步骤如下：

（1）鼠标左键点击"TCP Server"选中它，然后点击"创建"按钮，在弹出的对话框中输入端口号，比如"8080"，如图 5.2-4 所示。

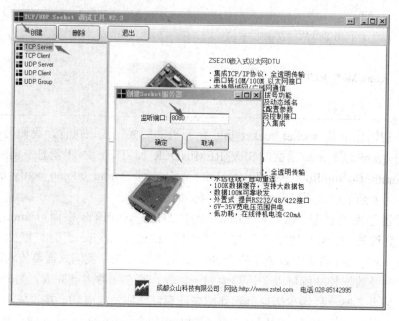

图 5.2-4

单击"确定"按钮之后，服务器就已经创建好了，服务器建好之后就会立即启动监听，等待终端连接，如图 5.2-5 所示，其中服务器的 IP 地址就是我们电脑上的网卡 IP 地址，不同的电脑，这个地址是不一样的，笔者的电脑 IP 地址为："172.20.65.107"

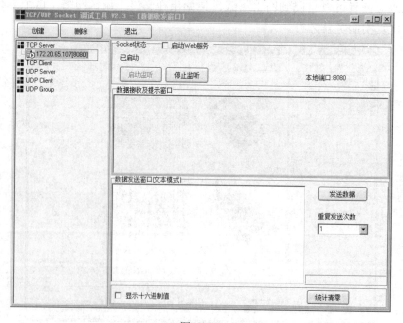

图 5.2-5

(2) 在 Socket.h 文件中设置我们的 Socket 要连接的服务器 IP 地址，修改 NETWORK_DEFAULT_SERVER_ADDR 宏定义的值为"172.20.65.107"，NETWORK_DEFAULT_SERVER_PORT 宏定义的值为 8080，运行模拟器，就可以在"TCP/UDP Socket 调试工具"中看到每隔一分钟就能收到一条数据，数据内容为"mystudy socket"加日期时间，如图 5.2-6 所示。

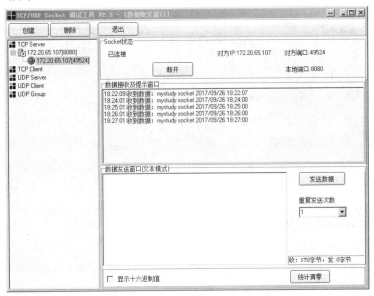

图 5.2-6

(3) 在 mmi_mystudy_socket_cb 函数中，给 case SOC_RECV_SUCCESS 语句中打断点(快捷键 F9)，然后在 TCP/UDP Socket 调试工具的数据发送窗口中输入"mystudy socket recv test"，点击"发送数据"，如图 5.2-7 所示。

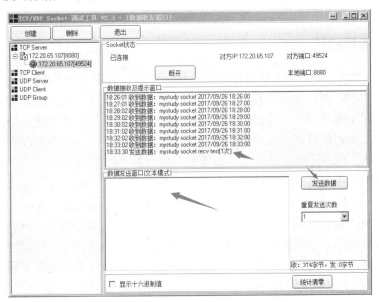

图 5.2-7

数据发送后，VS2008 中会立即执行到 mmi_mystudy_socket_cb 函数，并响应断点，接收的数据内容就保存在 soc_data->recv_data_buff 变量中，其内容就是我们在 TCP/UDP Socket 调试工具中发送的数据，如图 5.2-8 所示。

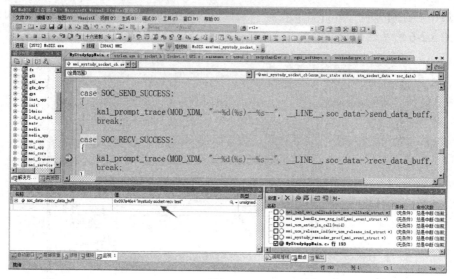

图 5.2-8

5.3 网络通信协议

网络通信协议有两种含义：一种是指网络连接协议，约定终端设备与服务器之间如何建立网络连接，这种协议都是行业标准规范；另一种是指网络数据传输协议，约定终端设备与服务器之间的数据传输格式，这种协议是我们自己定义的。

5.3.1 网络连接协议

在 MTK 智能设备研发中，目前最常用的网络连接协议是 TCP/IP 和 HTTP(超文本传输协议)，其中 TCP/IP 默认为长连接，长连接是指网络连接建立成功后能够维持较长的一段时间，如果期间不断地有数据收发，那么网络连接会一直维持下去而不会断开；HTTP 通常用作短连接，短连接只能由客户端主动往服务器发送数据，而服务器收到客户端的数据都必须回复，然后网络连接就断开了，短连接只能维持一次会话。

长连接和短连接通过代码实现的差别就在于数据的格式不同。比如发送 "hello mtk" 到服务器 www.baidu.com (百度域名)，其网址为http://www.baidu.com/index.html，假设服务器端口号为 8080，那么长连接发送的数据就是 "hello mtk"，不带其他任何格式修饰，但是短连接就稍微复杂些，它需要在 TCP/IP 连接的基础上附带 HTTP 协议的数据格式，HTTP 协议又分为多种，我们只介绍最常用的 GET 和 POST，数据格式分别如下所示：

(1) GET 格式，代码如下所示：

GET /index?=hello mtk   HTTP/1.1\r\n

Host: www.baidu.com:8080\r\n

Accept:*/*\r\n

Connection:close\r\n

User-Agent: Maui Browser\r\n\r\n

第一行"GET /index?=hello mtk　HTTP/1.1\r\n"：GET 是数据包的头，表示当前数据包格式为 GET 格式；"/index"是 URL，指向服务器 www.baidu.com 上的一个网页，"index"就是"index.html"；"?=hello mtk"中的"hello mtk"就是要上传到 URL(index.html)地址中的数据，"?="后面的内容全部是要上传的数据，这个数据格式必须符合 URL 编码，否则服务器接收到的内容与终端发送的内容不一致；"HTTP/1.1"表示 HTTP 协议的版本为 1.1。每一行的末尾都以一个"\r\n"结束，最后一行以两个"\r\n"结束。

第二行"Host: www.baidu.com:8080\r\n"："Host:"是固定字段，后面的"www.baidu.com:8080"表示服务器域名和端口号。

第三行"Accept:*/*\r\n"："Accept:"代表发送端(客户端)希望接受的数据类型，"*/*"表示任意数据类型。

第四行"Connection:close\r\n"："Connection:"表示连接属性，close 为短连接，keep-alive 为长连接，但是在嵌入式中，我们不推荐使用 HTTP 保持长连接。

第五行"User-Agent: Maui Browser\r\n\r\n"："User-Agent:"标志当前请求来自哪个用户，在 MTK 系统中可以使用 applib\misc\include\ app_ua.c 文件中的 applib_inet_get_user_agent_string 来获取该值。

HTTP GET 主要用于从服务器上获取数据，可以在 URL 中附带一些简单的参数信息，以上的 GET 数据包相当于把"hello mtk"字符传入到 http://www.baidu.com/index.html 网页中，我们可以直接把它拼成一个网址"http://www.baidu.com/index.html?= hello mtk"放到浏览器地址栏中，就可以把数据上传到服务器，并在浏览器中查看到服务器返回的结果。

在 Socket.h 文件中，我们已经把 GET 数据包格式定义成了一个宏定义 HTTP_GET_DATA_FORMAT，可以通过 sprintf 函数来格式化获取数据包，语句如下，得到的结果如图 5.3-1 所示。

sprintf(http_get, HTTP_GET_DATA_FORMAT, "/index", "?=hello mtk", "www.baidu.com", 8080, "Maui Browser");

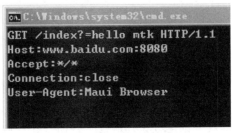

图 5.3-1

(2) POST 格式，代码如下所示：

POST /index　HTTP/1.1\r\n

Host: www.baidu.com:8080\r\n

Accept:*/*\r\n

　　　　Connection:close\r\n
　　　　User-Agent: Maui Browser\r\n
　　　　Content-Length:9\r\n\r\n
　　　　hello mtk

　　POST 主要用于把数据上传到服务器中，其数据包格式与 GET 类似，在第一行的 URL（"/index"）后面也可使用"?="添加参数，这样前面五行就与 GET 完全一样了。但是 URL 中能够携带的数据大小有限制，如果要上传的数据达到 1G 甚至更大，就无法通过 URL 上传了，只能用 POST 上传。在上述 POST 数据包中，第六行"Content-Length"表示要上传的数据长度，第七行开始就是需要上传的数据内容，这个数据内容的大小没有限制，但是长度必须等于"Content-Length"标识的长度。

　　在 Socket.h 文件中，POST 数据包格式也定义成了宏 HTTP_POST_DATA_FORMAT，通过 sprintf 函数格式化的语句如下，得到的结果如图 5.3-2 所示。

　　sprintf(http_post, HTTP_POST_DATA_FORMAT, "/index", "", "www.baidu.com", 8080, "Maui Browser", strlen("hello mtk"), "hello mtk");

图 5.3-2

　　在 MTK 设备开发中，我们主要使用的是 TCP/IP 长连接，因为只有在长连接的情况下，服务器才可以主动下发指令到终端，这样可以保证数据的及时响应，而且 TCP/IP 的数据传输有效率大于 HTTP，比如总共发送 9 字节的数据"hello MTK"，使用 HTTP 协议却需要加一堆对我们没用的数据包头信息，即耗费系统资源，也浪费用户流量。

　　但 HTTP 也并不是完全没有用处，通常我们上传文件时可以使用 HTTP POST，而从服务器上下载文件可以使用 HTTP GET。

　　使用长连接和短连接相结合的方式，可以有效提高网络传输的效率，长连接用来建立消息通道，短连接用来建立数据通道。

5.3.2　数据传输协议

　　数据传输协议是终端与服务器之间的一种约定，比如终端上传"1"，服务器返回"1ok"；服务器下发"call10086"，则终端就直接拨打"10086"。这种数据协议完全可以由开发者自己定义，只要终端和服务器双方都可以识别就行了。但是在行业内也有一些比较实用的数据协议格式，这里给大家介绍两种：RTLV 和 JSON。

　　（1）RTLV 数据传输协议。

　　RTLV(Repeat-Type-Length-Value)数据传输协议是笔者自己发明的，并已申请了专利。其中 T 表示数据包的类型，也就是名称；V 表示数据段的值；L 表示数据段 V 的长度；R

表示重复，但只是重复 L 和 V 部分；其实应该命名为 TRLV，只是笔者觉得 RTLV 比较好听一点而已。通用格式表示为：

| T(type) | L(length) | V(value) | L(length) | V(value) | ... |

这样的数据格式，好处在于方便解析，而且整个数据包的长度不需要固定，能够避免无效数据的传输。比如，我们规定 T 和 L 各占 1 个字节，其中 L 表示 V 的长度，取值范围为 0～255，L=0 为整个数据包的结束符，L=255 表示数据段 V 为无效数据，并且缺省。解析的时候，只需要一个指针，就可以遍历整个数据包，并把所有的数据(V)提取出来。比如数据包："T[3]aaa[4]bbbb[5]ccccc"（中括号"[]"中的数字为 ASCII 数值），解析的时候，指针首先指向第一个字节就可以取出 type(T)；然后后移 1 个字节，得到第一个数据段的长度为 3，再把指针后移 1 个字节，并取出 3 个字节的数据，得到第一个数据段；然后指针继续后移 3 个字节，得到第二个数据的长度；依次类推直到取出的长度为 0，就表示该数据包解析完成了。C 语言算法及示例代码如代码清单 5.3-1 所示，读者可直接在 VS2008 中新建一个 WIN32 控制台应用程序，运行代码。

---代码清单 5.3-1---

```c
#include <stdio.h>
#include <stdlib.h>
#include <string.h>
#include <stdarg.h>

#define RTLV_TYPE_SIZE (1)/*type 的字节长度*/


/*****************************************************************
*功  能：制作 RTLV 格式字符串
*参  数：
            rtlv [out]   -- 保存生成的 RTLV 字符串
            num [in]     -- 可变参数的个数，从 type 开始计算，并包括 type
            type [in]    -- 数据包名称
*返回值：rtlv
*****************************************************************/
unsigned char *rtlv_make_data(unsigned char *rtlv, unsigned char num, unsigned char *type, ...)
{
    va_list arg_ptr;
    unsigned char *var=NULL, i=0;

    if(NULL==rtlv || 0==num || NULL==type)
    {
        return NULL;
    }
    sprintf((char*)rtlv, "%s", type);
```

```c
        va_start(arg_ptr, type);

            for (i=0; i<num-1; i++)
            {
                var = va_arg(arg_ptr, unsigned char *);
                if (NULL==var || '\0'==var[0])
                {
                    rtlv[strlen((char*)rtlv)] = 0xFF;
                }
                else
                {
                    rtlv[strlen((char*)rtlv)] = strlen((char*)var);
                    strcat((char*)rtlv, (char*)var);
                }
            }
            va_end(arg_ptr);

            return rtlv;
        }


/*****************************************************************
*功    能：解析 RTLV 格式字符串
*参    数：
            rtlv [in]    -- 保存生成的 RTLV 字符串
            num  [in]    -- 可变参数的个数，从 type 开始计算，并包括 type
            type [in]    -- 数据包名称
*返回值：rtlv
*****************************************************************/
unsigned char *rtlv_parse_data(unsigned char *rtlv, unsigned char num, unsigned char *type, ...)
{
    va_list arg_ptr;
    unsigned char *var=NULL, *value=NULL, length=0, i=0;

    if (NULL==rtlv || 0==num)
    {
        return NULL;
    }

    if(NULL != type)
    {
```

```c
            strncpy((char*)type, (char*)rtlv, RTLV_TYPE_SIZE);
    }
    value = rtlv+RTLV_TYPE_SIZE;
    va_start(arg_ptr, type);
    for (i=0; i<num-1; i++)
    {
        var = va_arg(arg_ptr, unsigned char *);
        length = value[0];
        value++;
        if (0xFF==length)
        {
            if (NULL != var)
            {
                var[0] = '\0';
            }
        }
        else
        {
            if (NULL != var)
            {
                strncpy((char*)var, (char*)value, length);
            }
            value += length;
        }
    }
    va_end(arg_ptr);
    return type;
}

int main(void)
{
    unsigned char rtlv[128]={0x00}, type[2]={0x00}, data1[16]={0x00}, data2[16]={0x00}, data3[16]={0x00};
    rtlv_make_data(rtlv, 4, (unsigned char*)"T", "aaa", "bbbb", "ccccc");
    printf("rtlv make: %s\n", rtlv);
    rtlv_parse_data(rtlv, 4, type, data1, data2, data3);
    printf("rtlv parse: %s, %s, %s, %s\n", type, data1, data2, data3);
}
```

---

(2) JSON 数据传输协议。

关于 JSON(JavaScript 对象标记语言)协议，网上能够找到很全面的讲解，本章节就不再赘述，希望读者能够自己通过网络学习。需要声明的是，JSON 原本是用于 JavaScript 语言，而 MTK 平台使用的是 C 语言编写，如果我们要在 MTK 平台上使用 JSON，就必须把 JSON 协议用 C 语言实现。所幸 JSON 的 C 语言版本网上可以找到现成的代码，我们只需要移植过来就可以用了(因代码太多，请读者到源码中阅读，相关文件 cJSON.c、cJSON.h)。以下是儿童定位智能手表的网络通信协议，使用的就是用的 JSON 格式。

表 5.3　儿童定位智能手表网络通信协议

指令功能名称	服务器发送指令	终端执行指令
屏幕测试	{"cmd":"test", "confirmCode":"230", "respVal":"-1"}	手表屏幕上显示特定内容
SIM 通话	{"cmd":"test", "confirmCode":"231", "respVal":"10086"}	手表拨打电话到 10086
SIM 短信	{"cmd":"test", "confirmCode":"232", "respVal":"10086"}	手表发送短信到 10086
GPS 测试	{"cmd":"test", "confirmCode":"233", "respVal":"-1"}	手表发送 GPS 经纬度短信到用户手机上
时间同步	{"cmd":"test", "confirmCode":"234", "respVal":"201712301230"}	设置手表的时间为 2017 年 12 月 30 日 12:30

以上网络指令都是比较简单的 JSON 数据，JSON 的特点是每个数据都有一个名称，格式为"{name：data}"，每个 JSON 数据都必须用大括号包含，其中 name 通常为字符串，data 可以是字符串、整型、浮点型、布尔型数据，也可以是嵌套的 JSON 数据，还可以是数组，数组以中括号"[]"包含。"name：data"可以包含任意多个，并且顺序可以随意。比如以上的指令中""cmd":"test""，"test"就是数据值，"cmd"就是数据值的名称。我们从上面挑选"SIM 通话"指令，在代码中测试一下，在 MyStudyAppMain.c 文件中修改代码如下(代码清单 5.3-2)。

---代码清单 5.3-2---
```
#include "cJSON.h"
/*省略部分代码*/
void mmi_mystudy_socket_cb(enum soc_state state, stu_socket_data *soc_data)
{
    cJSON *root=NULL;
    char number[32]={0x00}, *respVal = NULL, *confirmCode=NULL;
    /*省略部分代码*/
    switch(state)
    {
        /*省略部分代码*/
        case SOC_RECV_SUCCESS:
        {
            /*{"cmd":"test", "confirmCode":"231", "respVal":"10086"}*/
            root = cJSON_Parse((char *)soc_data->recv_data_buff);/*解析 json 数据*/
```

```
            if(NULL != root)
            {
                respVal = cJSON_GetObjectItem_string(root, "respVal");
                        /*获取名称为 respVal 的数据值*/
                if(NULL != respVal)
                {
                    mmi_asc_to_ucs2((CHAR*)number, (CHAR*)respVal);
                        /*把 ASC 编码转换为 UCS2 编码*/
                    MakeCall((CHAR*)number);/*拨打电话*/
                }
            }
            cJSON_Delete(root);
            break;
        }
    }
}
```

以上代码中头文件"cJSON.h"包含了所有的 JSON 相关函数接口，请读者自己在源代码中下载该文件。JSON 数据的解析，首先要调用 cJSON *cJSON_Parse(const char *value)函数，生成一个 JSON 对象(C 语言中其实没有对象的概念，此处指的是 cJSON 指针)，然后就可以从这个 JSON 对象中提取数据。其中，函数 char *cJSON_GetObjectItem_string (cJSON *object, const char *string)提取的是字符串类型的数据，我们提取"respVal"的值就可以得到电话号码 10086，然后执行拨号操作。JSON 对象使用完后需要调用 void cJSON_Delete(cJSON *c)函数释放，否则会引起内存泄漏。

运行模拟器和"TCP/UDP 调试工具"，当"TCP/UDP 调试工具"收到了模拟器发送的指令后，即视为网络已经连接上了，此时发送字符串"{"cmd":"test", "confirmCode": "231", "respVal":"10086"}"，就可以看到模拟器进入了拨号界面，拨打的电话号码为 10086，如图 5.3-3 所示。

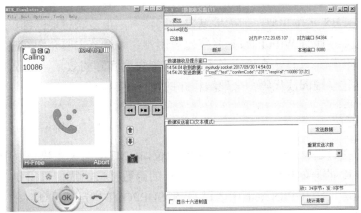

图 5.3-3

网络数据通信协议很多时候都是需要根据项目需求自己定义的。虽然有一些已经在行业内形成了规范，但也并不能适合所有项目的需求。笔者也是在这种情况下才发明了 RTLV 协议。协议的制定不仅要满足项目需求，还要方便编码和解析，并注重数据有效率。以上的 JSON 协议，其实传输效率并不高，因为它带了很多无用的数据，比如最简单的 JSON 数据 "{"A":0}"，这个数据占了 7 个字节，但只有 1 个字节(0)是对我们有用的，它的数据有效率约为 14%(1/7)。另外它只能应用于字符串，字符串的操作很耗资源。然而更糟糕的是，它在解析过程中还需要频繁地进行动态分配内存，这在嵌入式开发中属于禁忌，因为嵌入式设备的内存往往非常有限，如果频繁地动态分配内存很容易造成内存碎片，从而影响其他程序的运行，导致莫名其妙的死机现象。既然 JSON 协议这么多的弊端，为什么我们还要使用它呢？这是因为网络协议并非只在终端使用，我们还需要考虑服务器，服务器主要做数据处理及数据中转，比如我们的儿童定位智能手表，它的通信涉及三方：手表、服务器、微信客户端，而服务器和微信客户端之间只能使用 JSON，如果手表端不使用 JSON，则服务器需要作协议转换，这就增加了服务器的工作量及负荷。

5.4 定　　位

"定位"是指获取终端设备当前所在地球上的位置，并在地图上标记出来。它在智能穿戴和物联网设备中绝对属于非常重要的一个功能，基本上所有的智能穿戴和物联网设备都拥有定位功能，围绕着定位功能可以给用户提供很多外围服务，比如防丢、防盗、跟踪、运动轨迹、电子围栏等。实现定位的技术有很多种，但不管使用哪一种定位技术，最终在地图上标记位置都要使用经纬度，也就是说，定位的最终结果都是获取终端设备的经纬度。根据获取经纬度的方法，我们可以把所有的定位技术分为两类：终端定位和云端定位。终端定位是指终端设备直接获取自身的经纬度，这种定位方法目前只有 GPS 定位。云端定位是指服务器端通过一些运算方法，计算出终端设备的经纬度，这种定位方法一般有较大的误差，但可通过提高基础数据的准确性以及优化算法来提高定位计算的精准度，常见的有 LBS 定位和 WiFi 定位。

目前在智能穿戴和物联网设备中，比较常用的只有 GPS 定位和 LBS 定位，至于 WiFi 定位，其原理与 LBS 定位相似，但因地域限制比较大，很多农村地区都没有 WiFi 覆盖，所以使用的比较少，而且 WiFi 的数据收集比较困难。本章节主要介绍 GPS 定位和 LBS 定位，其中的示例代码，只有在真机设备上运行才能看到实际的效果，如果没有购买开发套件，读者只能通过详细阅读代码，来理解程序的运行逻辑。

5.4.1　GPS 定位

通常所说的卫星定位、北斗定位都是 GPS 定位。在 MTK 的大多数平台中都集成了 GPS 芯片，而且 MTK 的原始代码中也包含了 GPS 的函数接口，我们只需要在项目配置文件 make\HEXING60A_11B_GPRS.mak 中通过 GPS_SUPPORT 指定项目用到的 GPS 芯片即可，比如 GPS_SUPPORT=MT3336，这个值不能随便设置，一定要跟硬件上的 GPS 芯片一致，否则 GPS 无法工作。只要 GPS_SUPPORT 不等于 NONE，系统就会定义 GPS 的功能宏

"__GPS_SUPPORT__"，这个宏包含的代码就是 MTK 系统中原始的 GPS 代码。至于上层软件使用 GPS 定位其实很简单，主要分为三个步骤：打开 GPS、解析 GPS 数据、关闭 GPS。

为了方便代码管理，我们把定位功能单独作为一个 APP 开发，在 plutommi\MyStudyApp 目录下新建文件夹"Location"，然后在 plutommi\MyStudyApp\Location 目录下新建 Inc 和 Src 文件夹，分别再新建 gps.h 和 gps.c 文件，在 plutommi\mmi\Inc\MMI_features.h 文件中定义功能宏__MMI_MYSTUDY_LOCATION__，用于包含 GPS 相关的代码，最后要把 gps.c 和 gps.h 文件添加到 MyStudyApp 模块中参与编译。至于如何新增 APP，在前面已经做过很多示例，此处就不再详细介绍了。

GPS 相关的函数接口都定义在 plutommi\Service\MDI\MDISrc\mdi_gps.c 文件中，并且在 plutommi\MtkApp\EngineerMode\EngineerModeApp\EngineerModeAppSrc\EngineerMode-MiniGPS.c 文件中编写了完整的 GPS 示例代码。我们可以仿造它编写自己的 GPS 应用代码。

(1) 打开 GPS。

打开 GPS 需要做四步操作，首先使用 mdi_get_gps_port 函数获取 GPS 的端口号；再调用函数 mdi_gps_uart_open 以工作模式 MDI_GPS_UART_MODE_LOCATION 打开端口，然后使用 mdi_gps_set_work_port 函数设置 GPS 的工作端口；最后再次使用 mdi_gps_uart_open 函数设置端口工作模式为 MDI_GPS_UART_MODE_RAW_DATA。以上四个步骤必须按顺序执行，否则会有问题。在 gps.c 文件中编写代码如下(代码清单 5.4-1)：

--代码清单 5.4-1--

```c
void mmi_gps_open(void)
{
#if defined(__MTK_TARGET__)
    g_gps_handle.port = mdi_get_gps_port();
    if (g_gps_handle.port >= 0 && g_gps_handle.port_handle == -1)
    {
        g_gps_handle.port_handle = mdi_gps_uart_open((U16)g_gps_handle.port,
            MDI_GPS_UART_MODE_LOCATION, em_minigps_gps_callback);
        if (g_gps_handle.port_handle>=0 || MDI_RES_GPS_UART_ERR_PORT_ALREADY_
            OPEN==g_gps_handle.port_handle)
        {
            mdi_gps_set_work_port((U8)g_gps_handle.port);
        }
        g_gps_handle.port_handle = mdi_gps_uart_open((U16)g_gps_handle.port,
            MDI_GPS_UART_MODE_RAW_DATA, em_minigps_gps_callback);
    }
#else
    g_gps_handle.port_handle = 1;
#endif
}
```

这个函数无法单独运行通过(gps.c 和 gps.h 文件中的完整代码会在后面贴出来)，上面的代码中使用了一个宏"_MTK_TARGET_"，这个宏只有在真机设备中才会定义，如果在模拟器上运行，凡是被这个宏包含的代码都不会执行。模拟器上也有一个宏"WIN32"，凡是被这个宏包含的代码在真机上都不会执行。"_MTK_TARGET_"和"WIN32"这两个宏定义通常用于区别真机代码和模拟器代码。

mdi_gps_uart_open 函数的第一个参数是 GPS 的端口，第二个参数是 GPS 端口的工作模式，需要分两步分别设置为 MDI_GPS_UART_MODE_LOCATION 和 MDI_GPS_UART_MODE_RAW_DATA，最后一个参数是 GPS 端口的回调函数。

(2) 解析 GPS 数据。

当 GPS 开始工作后，就会不停地调用回调函数，并输出数据到回调函数中，我们可以在回调函数中解析 GPS 数据。为了方便读者阅读 EngineerModeMiniGPS.c 文件中的 GPS 代码，笔者把回调函数及相关的 GPS 数据解析函数定义为 EngineerModeMiniGPS.c 文件中的同名函数，并且功能也是一样的，只不过为了简化代码，部分不需要的函数内容空缺了。GPS 端口回调函数的定义代码如下(清单 5.4-2)：

---代码清单 5.4-2---

```
static void em_minigps_gps_callback(mdi_gps_parser_info_enum type, void *buffer, U32 length)
{
    switch(type)
    {
        case MDI_GPS_PARSER_NMEA_GGA:
            em_minigps_nmea_gga_callback(buffer);
            break;
        case MDI_GPS_PARSER_NMEA_RMC:
            em_minigps_nmea_rmc_callback(buffer);
            break;
        case MDI_GPS_PARSER_NMEA_GSA:
            em_minigps_nmea_gsa_callback(buffer);
            break;
        case MDI_GPS_PARSER_NMEA_GSV:
            em_minigps_nmea_gsv_callback(buffer);
            break;
        case MDI_GPS_PARSER_RAW_DATA:
            em_minigps_nmea_string_callback(buffer, length);
            break;
        case MDI_GPS_PARSER_NMEA_GAGSA:
            em_minigps_nmea_gagsa_callback(buffer);
            break;
        case MDI_GPS_PARSER_NMEA_GAGSV:
            em_minigps_nmea_gagsv_callback(buffer);
```

```
                    break;
                case MDI_GPS_PARSER_NMEA_GLGSA:
                    em_minigps_nmea_glgsa_callback(buffer);
                    break;
                case MDI_GPS_PARSER_NMEA_GLGSV:
                    em_minigps_nmea_glgsv_callback(buffer);
                    break;
            }
        }
```

GPS 数据采用的是 NMEA 数据协议，其中包括 GPGGA、GPRMC 等不同的 GPS 数据类型，我们在回调函数中使用 switch case 分支语句，分别处理不同的 GPS 数据。关于 NMEA 读者可以上网了解一下，这是行业内的通用标准，网上很容易找到专业的解释，这里就不多作介绍了。另外函数内调用到的其他函数，会在后面一起贴出。

(3) 关闭 GPS。

GPS 工作的时候比较耗电，所以为了节省功耗，在不需要 GPS 的时候就要把它关闭。关闭 GPS 需要调用函数 mdi_gps_uart_close，其参数传递与 mdi_gps_uart_open 函数传递的参数一模一样。代码如下(代码清单 5.4-3)：

--代码清单 5.4-3--

```
        void mmi_gps_close(void)
        {
            if(g_gps_handle.port_handle != -1)
            {
        #if defined(__MTK_TARGET__)
                mdi_gps_uart_close(g_gps_handle.port, MDI_GPS_UART_MODE_LOCATION,
                        em_minigps_gps_callback);
                mdi_gps_uart_close(g_gps_handle.port, MDI_GPS_UART_MODE_RAW_DATA,
                        em_minigps_gps_callback);
        #endif
                g_gps_handle.port_handle = -1;
                memset(&g_gps_data, 0x00, sizeof(g_gps_data));
            }
        }
```

GPS 定位不仅仅可以获取设备所在的经纬度，还可以获取设备的移动速度、移动方向、海拔高度等。但是 GPS 定位无法在室内使用，必须在室外无遮挡的地方才能有比较好的定位效果。如果周围有高楼大厦或者天气不好有云层遮挡，都有可能影响 GPS 的定位精准度。另外，硬件的设计也会直接影响 GPS 的定位效果，比如 GPS 天线性能，不仅影响定位精准度，还影响定位的速度。gps.h 和 gps.c 文件中的完整代码如下(代码清单 5.4-4 和 5.4-5)：

---------------------------------------gps.h 代码清单 5.4-4---------------------------------------

```c
#ifndef __GPS_H__
#define __GPS_H__

#include "mmi_features.h"

#if defined(__MMI_MYSTUDY_LOCATION__)&&defined(__GPS_SUPPORT__)

#include "MMIDataType.h"

/**********************************************************************
* 数据类型
**********************************************************************/

typedef struct
{
    FLOAT    longitude;      /*经度*/
    FLOAT    latitude;       /*纬度*/
    S8       north_south;    /*北-南*/
    S8       east_west;      /*东-西*/
}stu_gps_data;

typedef struct
{
    S32 port;         /*GPS 工作端口*/
    S32 port_handle;  /*-1:GPS 关闭状态; 否则 GPS 处于打开状态*/
}stu_gps_handle;


/**********************************************************************
* 函数接口
**********************************************************************/


/**********************************************************************
* 功  能: 获取 GPS 数据
* 参  数: void
* 返回值: GPS 数据
**********************************************************************/
extern stu_gps_data *mmi_gps_get_data(void);
```

```c
/*********************************************************************
 * 功    能：打开 GPS
 * 参    数：void
 * 返回值：void
 *********************************************************************/
extern void mmi_gps_open(void);


/*********************************************************************
 * 功    能：关闭 GPS
 * 参    数：void
 * 返回值：void
 *********************************************************************/
extern void mmi_gps_close(void);

#endif
#endif/*__GPS_H__*/
```

---gps.c-代码清单 5.4-5---

```c
#include "mmi_features.h"

#if defined(_MMI_MYSTUDY_LOCATION_)&&defined(_GPS_SUPPORT_)

/*********************************************************************
 * 头文件
 *********************************************************************/


#include "GPS.h"
#include "mdi_gps.h"
#include "gps_common.h"
/*********************************************************************
/* 宏常量定义
 *********************************************************************/


/*********************************************************************
/* 全局变量
 *********************************************************************/
static stu_gps_handle g_gps_handle = {0, -1};
static stu_gps_data g_gps_data = {0x00};
```

```c
/**********************************************************************
/* 函数声明
**********************************************************************/

/**********************************************************************
/* 函数体
**********************************************************************/

static void em_minigps_nmea_gga_callback(mdi_gps_nmea_gga_struct *param)
{
    /*----------------------------------------------------------------*/
    /* Local Variables                                                */
    /*----------------------------------------------------------------*/
    mdi_gps_nmea_gga_struct *gga_data = param;


    /*----------------------------------------------------------------*/
    /* Code Body                                                      */
    /*----------------------------------------------------------------*/
}

static void em_minigps_nmea_rmc_callback(mdi_gps_nmea_rmc_struct *param)
{
    /*----------------------------------------------------------------*/
    /* Local Variables                                                */
    /*----------------------------------------------------------------*/
    mdi_gps_nmea_rmc_struct *rmc_data = param;
    U8 trace_data[128]={0x00};
    /*----------------------------------------------------------------*/
    /* Code Body                                                      */
    /*----------------------------------------------------------------*/
    g_gps_data.latitude = rmc_data->latitude;
    g_gps_data.longitude = rmc_data->longitude;
    g_gps_data.north_south = rmc_data->north_south;
    g_gps_data.east_west = rmc_data->east_west;

    sprintf(trace_data, "lat:%f, lng:%f", g_gps_data.latitude, g_gps_data.longitude);
    kal_prompt_trace(MOD_XDM, "-gps-%d(%s)--%s--", __LINE__, (U8*)trace_data, __FILE__);
}
```

```
static void em_minigps_nmea_gsa_callback(mdi_gps_nmea_gsa_struct *param)
{
    /*----------------------------------------------------------------*/
    /* Local Variables                                                */
    /*----------------------------------------------------------------*/
    mdi_gps_nmea_gsa_struct *gsa_data = param;

    /*----------------------------------------------------------------*/
    /* Code Body                                                      */
    /*----------------------------------------------------------------*/
}

static void em_minigps_nmea_gsv_callback(mdi_gps_nmea_gsv_struct *param)
{
    /*----------------------------------------------------------------*/
    /* Local Variables                                                */
    /*----------------------------------------------------------------*/
    mdi_gps_nmea_gsv_struct *gsv_data = param;

    /*----------------------------------------------------------------*/
    /* Code Body                                                      */
    /*----------------------------------------------------------------*/
}

static void em_minigps_nmea_string_callback(const U8 *buffer, U32 length)
{
    /*----------------------------------------------------------------*/
    /* Local Variables                                                */
    /*----------------------------------------------------------------*/

    /*----------------------------------------------------------------*/
    /* Code Body                                                      */
    /*----------------------------------------------------------------*/

}

static void em_minigps_nmea_gagsa_callback(void *param)
{
```

```c
    /*----------------------------------------------------------------*/
    /* Local Variables                                                */
    /*----------------------------------------------------------------*/
    gps_common_nmea_gsa_struct *gsa_data = (gps_common_nmea_gsa_struct*)param;

    /*----------------------------------------------------------------*/
    /* Code Body                                                      */
    /*----------------------------------------------------------------*/

}

static void em_minigps_nmea_gagsv_callback(void *param)
{
    /*----------------------------------------------------------------*/
    /* Local Variables                                                */
    /*----------------------------------------------------------------*/
    gps_common_nmea_gsv_struct *gsv_data = (gps_common_nmea_gsv_struct*)param;

    /*----------------------------------------------------------------*/
    /* Code Body                                                      */
    /*----------------------------------------------------------------*/

}

static void em_minigps_nmea_glgsa_callback(void *param)
{
    /*----------------------------------------------------------------*/
    /* Local Variables                                                */
    /*----------------------------------------------------------------*/
    gps_common_nmea_gsa_struct *gsa_data = (gps_common_nmea_gsa_struct*)param;

    /*----------------------------------------------------------------*/
    /* Code Body                                                      */
    /*----------------------------------------------------------------*/

}

static void em_minigps_nmea_glgsv_callback(void *param)
{
    /*----------------------------------------------------------------*/
```

```
    /* Local Variables                                          */
    /*------------------------------------------------------------*/
    gps_common_nmea_gsv_struct *gsv_data = (gps_common_nmea_gsv_struct*)param;

    /*------------------------------------------------------------*/
    /* Code Body                                                 */
    /*------------------------------------------------------------*/
}

static void em_minigps_gps_callback(mdi_gps_parser_info_enum type, void *buffer, U32 length)
{
    kal_prompt_trace(MOD_XDM, "--%d(gps:%s)--%s--", __LINE__, (U8*)buffer, __FILE__);
    switch(type)
    {
        case MDI_GPS_PARSER_NMEA_GGA:
            em_minigps_nmea_gga_callback(buffer);
            break;
        case MDI_GPS_PARSER_NMEA_RMC:
            em_minigps_nmea_rmc_callback(buffer);
            break;
        case MDI_GPS_PARSER_NMEA_GSA:
            em_minigps_nmea_gsa_callback(buffer);
            break;
        case MDI_GPS_PARSER_NMEA_GSV:
            em_minigps_nmea_gsv_callback(buffer);
            break;
        case MDI_GPS_PARSER_RAW_DATA:
            em_minigps_nmea_string_callback(buffer, length);
            break;
        case MDI_GPS_PARSER_NMEA_GAGSA:
            em_minigps_nmea_gagsa_callback(buffer);
            break;
        case MDI_GPS_PARSER_NMEA_GAGSV:
            em_minigps_nmea_gagsv_callback(buffer);
            break;
        case MDI_GPS_PARSER_NMEA_GLGSA:
            em_minigps_nmea_glgsa_callback(buffer);
            break;
        case MDI_GPS_PARSER_NMEA_GLGSV:
```

```
                    em_minigps_nmea_glgsv_callback(buffer);
                    break;
            }
    }

    void mmi_gps_open(void)
    {
    #if defined(__MTK_TARGET__)
        g_gps_handle.port = mdi_get_gps_port();
        if (g_gps_handle.port >= 0 && g_gps_handle.port_handle == -1)
            {
                g_gps_handle.port_handle = mdi_gps_uart_open((U16)g_gps_handle.port,
                        MDI_GPS_UART_MODE_LOCATION, em_minigps_gps_callback);
                if (g_gps_handle.port_handle>=0 || MDI_RES_GPS_UART_ERR_PORT_ALREADY_
                        OPEN==g_gps_handle.port_handle)
                {
                    mdi_gps_set_work_port((U8)g_gps_handle.port);
                }
                g_gps_handle.port_handle = mdi_gps_uart_open((U16)g_gps_handle.port,
                        MDI_GPS_UART_MODE_RAW_DATA, em_minigps_gps_callback);
            }
    #else
        g_gps_handle.port_handle = 1;
    #endif
        kal_prompt_trace(MOD_XDM, "port:%d handle:%d", g_gps_handle.port,
                    g_gps_handle.port_handle);
    }

    /***********************************************************************
    * 功  能：关闭 GPS
    * 参  数：
    *        void
    * 返回值：
    *        void
    ***********************************************************************/
    void mmi_gps_close(void)
    {
        /*----------------------------------------------------------------*/
        /* Local Variables                                            */
```

```c
/*--------------------------------------------------------------*/

/*--------------------------------------------------------------*/
/* Code Body                                                    */
/*--------------------------------------------------------------*/
    if(g_gps_handle.port_handle != -1)
    {
        kal_prompt_trace(MOD_XDM, "--mmi_gps_close--%d-%s-", __LINE__, __FILE__);
#if defined(__MTK_TARGET__)
        mdi_gps_uart_close(g_gps_handle.port, MDI_GPS_UART_MODE_LOCATION,
                em_minigps_gps_callback);
        mdi_gps_uart_close(g_gps_handle.port, MDI_GPS_UART_MODE_RAW_DATA,
                em_minigps_gps_callback);
#endif
        g_gps_handle.port_handle = -1;
        memset(&g_gps_data, 0x00, sizeof(g_gps_data));
    }
}

stu_gps_data *mmi_gps_get_data(void)
{
    return &g_gps_data;
}
#endif/*defined(__MMI_MYSTUDY_LOCATION__)*/
```

接下来,我们在设备一开机的时候,就打开 GPS,在 MyStudyAppMain.c 文件末尾处添加函数 mmi_mystudy_app_init,并在函数内调用 mmi_gps_open 打开 GPS,代码如下(代码清单 5.4-6):

--代码清单 5.4-6--

```c
#include "GPS.h"

/*省略部分代码*/
void mmi_mystudy_app_init(void)
{
    mmi_gps_open();/*打开 GPS*/
}
#endif
```

----

mmi_mystudy_app_init 函数作为 MystudyApp 模块的初始化函数,后面所有开机就需要

执行的操作都放在这个函数中执行。系统所有开机操作完成后，会执行 plutommi\mmi\Bootup\BootupSrc\BootupInitApps.c 文件中的 mmi_bootup_notify_completed 函数，我们把 mmi_mystudy_app_init 放到这个函数中调用就可以了。代码如下(代码清单 5.4-7)：

--代码清单 5.4-7--

```
/*省略系统默认代码*/
#if defined(__MYSTUDY_APPLICATION__)
extern void mmi_mystudy_app_init(void);
#endif
void mmi_bootup_notify_completed(void)
{
    /*省略系统默认代码*/
    #if defined(__MYSTUDY_APPLICATION__)
        mmi_mystudy_app_init();
    #endif
}
```

因为新增了源文件，所以需要使用 make new 指令编译代码，并且模拟器也要使用 make gen_moids 重新生成。make new 指令编译通过后，使用 Flash Tools 工具把系统烧录到真机设备中，开机之后连上 Catcher 就可以查看 GPS 的工作情况了。我们在函数 em_minigps_gps_callback 中打印了一行 log，不断地输出 GPS 数据，如果可以看到如图 5.4-1 所示的内容，则说明 GPS 启动成功。

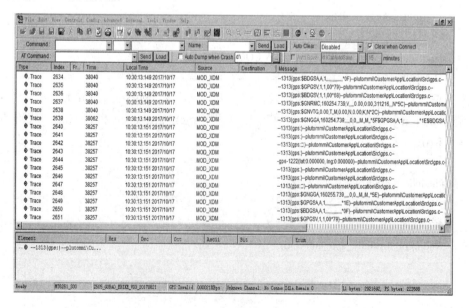

图 5.4-1

以上的打印出来的 log 信息，都是无效的 GPS 数据，因为笔者是在室内测试的。如果要获取有效的 GPS 数据，必须拿到室外去测试。为了方便室外测试，我们再增加一个功能，每隔一分钟获取一次 GPS 数据，并以短信的方式发送到自己的手机上，如果成功获取经纬度，则关闭 GPS。在 MyStudyAppMain.c 文件的 mmi_mystudy_reminder_proc 函数中编写代码如下(代码清单 5.4-8)：

---代码清单 5.4-8---

```c
mmi_ret mmi_mystudy_reminder_proc(mmi_event_struct *evt)
{
    /*省略部分代码*/
    U8 soc_data[64] = {0x00}, location_data[256]={0x00}, temp[128]={0x00};
    stu_gps_data *gps_data=NULL;

    /*省略部分代码*/

    switch (reminder_evt->notify)
    {
        /*省略部分代码*/
        case SRV_REMINDER_NOTIFY_EXPIRY:/*触发*/
        {
            /*省略部分代码*/
            gps_data = mmi_gps_get_data();
            sprintf((char*)temp, "gps(%f, %f)", gps_data->latitude, gps_data->longitude);
            mmi_asc_to_ucs2((S8*)location_data, (S8*)temp);
            srv_sms_send_ucs2_text_msg((char*)location_data, (char*)L"18900000000",
                        SRV_SMS_SIM_1, mmi_send_sms_callback, NULL);
            if(0!=gps_data->latitude*10000 && 0!=gps_data->longitude*10000)
            {
                mmi_gps_close();
            }

            mmi_set_reminder_time();

            break;
        }
        /*省略部分代码*/
    }
    return MMI_RET_OK;
}
```

经度的取值范围为 0～180°，分为东经(E)和西经(W)，纬度的取值范围为 0～90°，分为南纬(S)和北纬(N)。系统获取的经纬度默认都精确到六位小数，在判断经纬度是否有效时，

笔者把它放大了 10000 倍，因为存储经纬度的变量默认值都为 0，如果放大 10000 依旧为 0，则判断为无效数据。另外如果我们刚好站在地球上经纬度都为 0 的本初子午线和赤道交汇的点上，则也被判断为无效数据，但这种情况我们可以不用考虑，因为没有读者会在非洲大西洋上验证本书上的 GPS 定位代码。把上述代码中发送短信的语句中的电话号码"18900000000"替换成读者自己的电话号码，使用 make r mystudyapp 编译代码，自己的手机上将会显示短信内容如图 5.4-2 所示。

图 5.4-2

获取到经纬度后，可以在 http://www.gpsspg.com/maps.htm 在线地图上查看经纬度所在的位置。笔者搜到的经纬度坐标为(22.525008, 113.997830)，在地图上显示的位置如图 5.4-3 所示，正是笔者测试所在的位置。需要注意的是，在地图页面左下侧需要选择"硬件/谷歌地球卫星"，否则显示的位置会有差异。

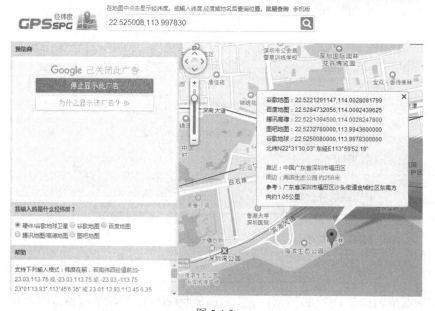

图 5.4-3

5.4.2 LBS 定位

LBS(Location Based Service)定位也称之为基站定位,它通过 SIM 卡把接收的基站信息上传到服务器,服务器端从数据库中查找到对应的基站位置信息,再通过几何运算计算出终端设备的具体位置。所谓的基站就是指电信运营商的信号塔,在中国主要有三大运营商:移动、联通、电信,其中移动的基站覆盖面积最广、密度最大,所以基于移动 SIM 卡的 LBS 定位精准度更高。因为 LBS 定位是在服务器端计算出经纬度,所以称之为云端定位,其原理就是通过基站的经纬度位置及信号衰减来判断终端的位置。比如我们的手机收到了基站 A 发出的信号,假设基站 A 信号覆盖面积为 20 公里,那么就可以判断出手机处在距离基站 A 最大 20 公里的范围内,否则手机就无法接收到基站 A 的信号。另外,基站的信号强度会根据距离变化而变化,距离越远,信号越弱,根据距离与信号强度的衰减规律,可以判断手机与基站 A 的大概距离。假设手机同时还接收到了其他基站的信号,比如基站 B、基站 C,我们同样可以判断出手机与基站 B、基站 C 的距离。如果已知基站 A、基站 B、基站 C 的经纬度坐标,并且三个基站不在同一条直线上,那就可以确定手机的具体位置,这就是多基站定位的大致原理,如图 5.4-4 所示。

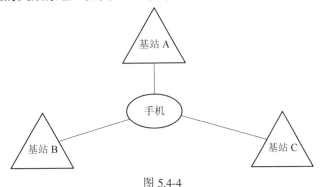

图 5.4-4

实际上 LBS 定位要比前面描述的复杂很多,只要一个基站信息也可以确定用户的位置,只是误差会比较大。LBS 定位依赖于基站的位置信息,在服务器端必须有记录了所有基站位置信息的数据库,这个数据库掌握在运营商手里,需要花钱购买,也可以直接使用一些电子地图提供的 LBS API,这样可以直接使用第三方的算法,像百度、高德都有这方面的服务。这些都是服务器要做的工作,我们只需要简单了解其原理就行了。至于终端要做的事情,就是把获取到的基站信息上传给服务器就可以了。

LBS 定位不需要打开和关闭操作,它主要是获取基站信息。只要 SIM 卡能搜到网络,设备就一定能够获取到基站信息,我们只需要监听 MSG_ID_L4C_NBR_CELL_INFO_REG_CNF 和 MSG_ID_L4C_NBR_CELL_INFO_IND 消息 id 就可以了,在监听函数里面提取出基站信息,代码如下(代码清单 5.4-9):

------------------------------------代码清单 5.4-9------------------------------------
```
static void mmi_lbs_cell_info(l4c_nbr_cell_info_ind_struct *msg_ptr)
{
    gas_nbr_cell_info_struct cell_info={0x00};
```

```c
U8 i = 0;

if(NULL == msg_ptr)
{
    return;
}
if (KAL_TRUE == msg_ptr->is_nbr_info_valid)
{
    memcpy((void *)&cell_info, (void *)(&(msg_ptr->ps_nbr_cell_info_union.gas_nbr_
            cell_info)), sizeof(gas_nbr_cell_info_struct));
}
else
{
    memset((void *)&cell_info, 0x00, sizeof(gas_nbr_cell_info_struct));
}

if(cell_info.nbr_cell_num <= 0)
{
    return;
}

memset(&g_lbs_data, 0x00, sizeof(g_lbs_data));
g_lbs_data.count = cell_info.nbr_cell_num+1;         /*搜索到的基站个数*/
if(g_lbs_data.count > LBS_SAL_CELL_NBR_MAX)
{
    g_lbs_data.count = LBS_SAL_CELL_NBR_MAX;
}
g_lbs_data.info[0].arfcn = cell_info.nbr_meas_rslt.nbr_cells
                    [cell_info.serv_info.nbr_meas_rslt_index].arfcn;
g_lbs_data.info[0].bsic = cell_info.nbr_meas_rslt.nbr_cells[cell_info.serv_info.nbr_meas_
                    rslt_index].bsic;
g_lbs_data.info[0].rxlev = cell_info.nbr_meas_rslt.nbr_cells[cell_info.serv_info.nbr_meas_
                    rslt_index].rxlev;
g_lbs_data.info[0].mcc = cell_info.serv_info.gci.mcc;
g_lbs_data.info[0].mnc = cell_info.serv_info.gci.mnc;
g_lbs_data.info[0].lac = cell_info.serv_info.gci.lac;
g_lbs_data.info[0].ci = cell_info.serv_info.gci.ci;

for(i = 0; i<cell_info.nbr_cell_num && i<LBS_SAL_CELL_NBR_MAX-1; i++)
```

```c
        {
            g_lbs_data.info[i+1].arfcn = cell_info.nbr_meas_rslt.nbr_cells[cell_info.nbr_cell_
                            info[i].nbr_meas_rslt_index].arfcn;
            g_lbs_data.info[i+1].bsic = cell_info.nbr_meas_rslt.nbr_cells[cell_info.nbr_cell_
                            info[i].nbr_meas_rslt_index].bsic;
            g_lbs_data.info[i+1].rxlev = cell_info.nbr_meas_rslt.nbr_cells[cell_info.nbr_cell_
                            info[i].nbr_meas_rslt_index].rxlev;
            g_lbs_data.info[i+1].mcc = cell_info.nbr_cell_info[i].gci.mcc;
            g_lbs_data.info[i+1].mnc = cell_info.nbr_cell_info[i].gci.mnc;
            g_lbs_data.info[i+1].lac = cell_info.nbr_cell_info[i].gci.lac;
            g_lbs_data.info[i+1].ci = cell_info.nbr_cell_info[i].gci.ci;
        }
    }

    void mmi_lbs_update(void)
    {
        SetProtocolEventHandler(mmi_lbs_cell_info, MSG_ID_L4C_NBR_CELL_INFO_REG_CNF);
        SetProtocolEventHandler(mmi_lbs_cell_info, MSG_ID_L4C_NBR_CELL_INFO_IND);
        mmi_frm_send_ilm(MOD_L4C, MSG_ID_L4C_NBR_CELL_INFO_REG_REQ, NULL, NULL);
    }
```

以上代码编写在 lbs.c 文件中，请读者自己在 plutommi\MyStudyApp\Location 目录下对应的文件夹中新建 lbsh 和 lbs.c 文件，并把它们添加到 MyStudyApp.mak 模块中，其中 mmi_lbs_update 函数就相当于 LBS 的启动函数，mmi_lbs_cell_info 就是消息监听函数，在这个函数中获取 LBS 的数据信息，需要获取的信息请查看 LBS.h 文件中的结构体 lbs_cell_info_struct。LBS.C 和 LBS.H 文件的完整代码如下(代码清单 5.4-10)：

--lbs.h 代码清单 5.4-10--

```c
#ifndef __LBS_H__
#define __LBS_H__

#include "mmi_features.h"

#if defined(__MMI_MYSTUDY_LOCATION__)
#include "MMIDataType.h"

/***************************************************************
* 数据类型
***************************************************************/
#define LBS_SAL_CELL_NBR_MAX 6
```

```c
typedef struct{
    U16 arfcn;              /*ARFCN (绝对无线频道编号  Absolute Radio Frequency Channel
                              Number - ARFCN)*/
    U8  bsic;               /*BSIC (基站识别码  Base Station Identity Code )*/
    U8  rxlev;              /*Received signal level  信号强度等级*/
    U16 mcc;                /*MCC 移动国家码*/
    U16 mnc;                /*MNC 移动网络码*/
    U16 lac;                /*LAC 位置区码 location area code */
    U16 ci;                 /*CI 小区识别码 Cell ID*/
}lbs_cell_info_struct;

typedef struct{
    U16 count;
    lbs_cell_info_struct info[LBS_SAL_CELL_NBR_MAX];
}lbs_data_struct;


/************************************************************************
 * 函数声明
 ************************************************************************/

/*获取 LBS 信息*/
extern lbs_data_struct *mmi_lbs_get_data(void);
#endif


#endif/*__LBS_H__*/
```

---lbs.c 代码清单 5.4-11----------------------------------
```c
#include "mmi_features.h"


#if defined(__MMI_MYSTUDY_LOCATION__)
/***********************************************************
 * 头文件
 ***********************************************************/


#include "LBS.h"
#include "nbr_public_struct.h"
```

```
/***********************************************************
/* 宏常量定义
***********************************************************/


/***********************************************************
/* 全局变量
***********************************************************/
static lbs_data_struct g_lbs_data = {0x00};


/***********************************************************
/* 函数声明
***********************************************************/


/***********************************************************
/* 函数体
***********************************************************/
/*获取基站*/
static void mmi_lbs_cell_info(l4c_nbr_cell_info_ind_struct *msg_ptr)
{
    /*----------------------------------------------------------------*/
    /* Local Variables                                              */
    /*----------------------------------------------------------------*/
     gas_nbr_cell_info_struct cell_info={0x00};
     U8 i = 0;
    /*----------------------------------------------------------------*/
    /* Code Body                                                    */
    /*----------------------------------------------------------------*/
     if(NULL == msg_ptr)
     {
         return;
     }
     if (KAL_TRUE == msg_ptr->is_nbr_info_valid)
     {
         memcpy((void *)&cell_info, (void *)(&(msg_ptr->ps_nbr_cell_info_union.gas_
             nbr_cell_info)), sizeof(gas_nbr_cell_info_struct));
     }
     else
     {
```

```c
            memset((void *)&cell_info, 0x00, sizeof(gas_nbr_cell_info_struct));
    }

    kal_prompt_trace(MOD_WAP, "-lbs-%d(cell_num=%d)-%s-", _LINE_,
            cell_info.nbr_cell_num, _FILE_);
    if(cell_info.nbr_cell_num <= 0)
    {
            return;
    }

    memset(&g_lbs_data, 0x00, sizeof(g_lbs_data));
    g_lbs_data.count = cell_info.nbr_cell_num+1;    /*搜索到的基站个数*/
    if(g_lbs_data.count > LBS_SAL_CELL_NBR_MAX)
    {
        g_lbs_data.count = LBS_SAL_CELL_NBR_MAX;
    }
    g_lbs_data.info[0].arfcn = cell_info.nbr_meas_rslt.nbr_cells[cell_info.serv_info.nbr_meas_
                    rslt_index].arfcn;
    g_lbs_data.info[0].bsic = cell_info.nbr_meas_rslt.nbr_cells[cell_info.serv_info.nbr_meas_
                    rslt_index].bsic;
    g_lbs_data.info[0].rxlev = cell_info.nbr_meas_rslt.nbr_cells[cell_info.serv_info.nbr_meas_
                    rslt_index].rxlev;
    g_lbs_data.info[0].mcc = cell_info.serv_info.gci.mcc;
    g_lbs_data.info[0].mnc = cell_info.serv_info.gci.mnc;
    g_lbs_data.info[0].lac = cell_info.serv_info.gci.lac;
    g_lbs_data.info[0].ci = cell_info.serv_info.gci.ci;

    for(i = 0; i<cell_info.nbr_cell_num && i<LBS_SAL_CELL_NBR_MAX-1; i++)
    {
        g_lbs_data.info[i+1].arfcn = cell_info.nbr_meas_rslt.nbr_cells[cell_info.nbr_cell_
                            info[i].nbr_meas_rslt_index].arfcn;
        g_lbs_data.info[i+1].bsic = cell_info.nbr_meas_rslt.nbr_cells[cell_info.nbr_cell_
                            info[i].nbr_meas_rslt_index].bsic;
        g_lbs_data.info[i+1].rxlev = cell_info.nbr_meas_rslt.nbr_cells[cell_info.nbr_cell_
                            info[i].nbr_meas_rslt_index].rxlev;
        g_lbs_data.info[i+1].mcc = cell_info.nbr_cell_info[i].gci.mcc;
        g_lbs_data.info[i+1].mnc = cell_info.nbr_cell_info[i].gci.mnc;
        g_lbs_data.info[i+1].lac = cell_info.nbr_cell_info[i].gci.lac;
        g_lbs_data.info[i+1].ci = cell_info.nbr_cell_info[i].gci.ci;
```

```
        }

    #if 1
        for(i = 0; i<g_lbs_data.count; i++)
        {
            kal_prompt_trace(MOD_WAP, "-lbs-%d(mcc=%d, mnc=%d, lac=%d, ci=%d, rxlev=%d, bsic=%d, arfcn=%d)-%s-", __LINE__,
                g_lbs_data.info[i].mcc, g_lbs_data.info[i].mnc, g_lbs_data.info[i].lac, g_lbs_data.info[i].ci,
                g_lbs_data.info[i].rxlev, g_lbs_data.info[i].bsic, g_lbs_data.info[i].arfcn, __FILE__);
        }
    #endif
    }

    /*LBS 信息更新*/
    void mmi_lbs_update(void)
    {
        SetProtocolEventHandler(mmi_lbs_cell_info, MSG_ID_L4C_NBR_CELL_INFO_REG_CNF);
        SetProtocolEventHandler(mmi_lbs_cell_info, MSG_ID_L4C_NBR_CELL_INFO_IND);
        mmi_frm_send_ilm(MOD_L4C, MSG_ID_L4C_NBR_CELL_INFO_REG_REQ, NULL, NULL);
    }

    /*获取 LBS 信息*/
    lbs_data_struct *mmi_lbs_get_data(void)
    {
        return &g_lbs_data;
    }

    #endif/*__MMI_MYSTUDY_LOCATION__*/
```

为了区分 GPS 的 log 信息，我们使用 MOD_WAP 来打印 LBS 的数据，在 MyStudyAppMain.c 文件的函数 mmi_mystudy_app_init 中调用 mmi_lbs_update 函数启动 LBS 定位。代码如下(代码清单 5.4-12)：

--代码清单 5.4-12--

```
#include "LBS.h"

/*省略部分代码*/
void mmi_mystudy_app_init(void)
{
    mmi_gps_open();/*打开 GPS*/
```

```
        mmi_lbs_update()/*启动 LBS*/
    }
    #endif
```

使用 make r mystudyapp 编译代码，下载到真机上，开机后连接 Catcher 工具，注意在 Filter 中只选择 MOD_WAP，去掉 MOD_XDM，否则 log 信息太多。如果能够不断打印出 mmi_lbs_cell_info 函数中的 trace 信息，则说明系统正在获取 LBS 数据。如图 5.4-5 所示。

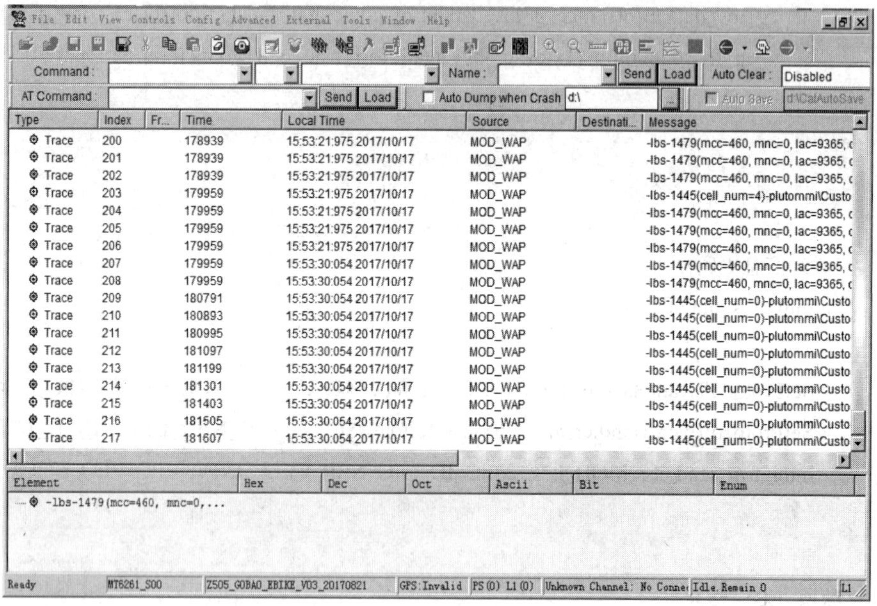

图 5.4-5

LBS 数据在室内也可以获取，我们没必要跑到户外去测试，可以在网页 http://www.gpsspg.com/bs.htm 中测试获取到的基站数据所定位的位置信息。

  LBS 定位和 GPS 定位各有各的优点和缺点。GPS 定位受周边环境影响较大，必须是室外无遮挡并且空旷的地方才能有比较好的定位效果，它需要在硬件上集成 GPS 芯片，需要消耗电能，对于一些功耗要求比较高的设备，我们要尽量减少它的启动次数和启动时间。而 LBS 定位只要在 SIM 卡有信号的情况下就能够定位，它不需要专门的芯片支持，也不需要考虑功耗问题，只不过定位精准度不及 GPS，因为 GPS 的发展已经很成熟了。但是随着技术的发展，LBS 定位也会越来越成熟，在一些大城市里面，多基站定位的精准度完全可以媲美 GPS，而且室内定位首选也是 LBS。在智能穿戴的开发中，我们通常都是 LBS 定位和 GPS 定位相结合使用。但一些物联网设备，比如摩拜单车，因为它们的应用场景为室外，通常只有 GPS 定位。

第六章 游戏开发

通过前面章节的学习之后，本章节笔者就带领大家开发一款小游戏。游戏开发趣味性比较强，可以提高读者的学习兴趣，而且涉及的知识面也比较广，涵盖了我们前面大部分章节中的知识点，包括界面绘制、层、定时器、动画播放、音乐播放等。另外还有一些我们前面没有讲到的知识点，比如触屏，这些知识点要么用得比较少，要么内容比较简单，笔者将在本章节结合游戏实例来给大家一一讲解。

本章的游戏实例，只需要在模拟器上运行即可，屏幕尺寸使用 128×128 的配置，另外还支持触屏功能，在下载的源码中提供了三款游戏：打地鼠、翻卡片、乐器演奏。我们以"打地鼠"游戏为例讲解，另外两个游戏供读者自己学习。如果还有对"打地鼠"游戏不熟悉的读者，可以先在网页上了解一下。

6.1 触　　屏

相信大家对触屏（即触摸屏）都不陌生，现在几乎所有的智能手机都是使用触屏操作的。在 MTK 系统中也支持触摸屏幕，只需在项目配置文件 make\HEXING61A_GPS_11C_GPRS.mak 中设置 TOUCH_PANEL_SUPPORT 宏的值不为 NONE 既可打开触屏功能，至于其值具体如何设置，在真机上需要根据硬件屏幕来配置，但在模拟器上可以随便设置。本例中，我们设置 TOUCH_PANEL_SUPPORT = CTP_FOCALTEK_FT5206_TRUE_MULTIPLE，再设置屏幕尺寸 MAIN_LCD_SIZE = 128×128。当使用 make 指令编译时，会报如下错误，如图 6.1-1 所示。

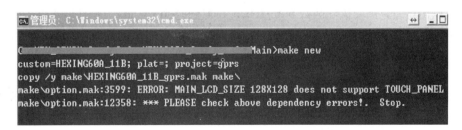

图 6.1-1

这个报错的地方在 make\option.mak 文件中的 3599 行，提示内容为"MAIN_LCD_SIZE 128X128 does not support TOUCH_PANEL"。原因是 MTK 系统中 128×128 的屏幕尺寸，默认不支持触屏功能，不过没有关系，我们在 option.mak 文件中把检测触屏和屏幕尺寸的地方注释掉就可以了。如图 6.1-2 所示。

```
03590:
03591: ifdef MAIN_LCD_SIZE
03592:   ifeq ($(call Upper,$(strip $(MAIN_LCD_SIZE))),120X160)
03593:     # spacial case for MTKLCM_COLOR in MT6208 EVB
03594:     MAIN_LCD_SIZE := 128X160
03595:   endif
03596:
03597: #  ifeq ($(call Upper,$(strip $(MAIN_LCD_SIZE))),128X128)
03598: #    ifneq ($(strip $(TOUCH_PANEL_SUPPORT)),NONE)
03599: #      $(warning ERROR: MAIN_LCD_SIZE 128X128 does not support TOUCH_PANEL)
03600: #      DEPENDENCY_CONFLICT = TRUE
03601: #    endif
03602: #  endif
03603:   COM_DEFS    += __MMI_MAINLCD_$(call Upper,$(strip $(MAIN_LCD_SIZE)))__
03604: endif
03605:
```

图 6.1-2

把 3597 行至 3602 行全部注释掉，再使用 make new 编译，就不会再报错了，而且 128×128 的屏幕尺寸也可以支持触屏功能。触屏功能的使用非常简单，它分为四个事件：按下、滑动、弹起、长按，其中长按事件用得比较少，而且有的 MTK 平台不支持，本书不作介绍。

(1) 触屏按下事件。

按下事件，是触屏操作响应的第一个事件，当我们点击屏幕时，它就会响应。可以使用函数 void wgui_register_pen_down_handler(mmi_pen_hdlr pen_down_hdlr)注册按下事件的响应函数。这个函数需要传如 typedef void (*mmi_pen_hdlr) (mmi_pen_point_struct pos)类型的函数指针。用法示例如代码清单 6.1-1 所示。

---代码清单 6.1-1--

void mmi_touch_test_pen_down_hdlr(mmi_pen_point_struct   pen_pos)

{

　　/*nothing*/

}

wgui_register_pen_down_handler (mmi_touch_test_pen_down_hdlr);

每次点击屏幕，都会执行函数 mmi_touch_test_pen_down_hdlr，并且传入点击的坐标位置，这个坐标位置我们通常都要保存起来。通过坐标值来判断该处理什么事情，一般情况下这个函数中只做一些按下的界面显示效果，比如按钮。

(2) 触屏的滑动事件。

当我们点击屏幕然后在屏幕上滑动时，就会响应滑动事件。注册滑动事件的函数接口为 void wgui_register_pen_move_handler(mmi_pen_hdlr pen_move_hdlr)，使用方法也是一样的。这个函数中，通常做一些拖动的界面显示效果。

(3) 触屏的弹起事件。

当我们点击屏幕然后松开的时候就会响应弹起事件。注册弹起事件的函数接口为 void wgui_register_pen_up_handler(mmi_pen_hdlr pen_up_hdlr)，使用方法同上。通常在这个函数中响应功能的入口函数。

在我们的游戏中，只需要判断触屏的"点击"和"滑动"事件就可以了，如果按下和弹起的坐标一样或小于某个距离，就判断为"点击"，否则就判断为"滑动"，然后比较按下和弹起的坐标值，判断滑动的方向。笔者将滑动事件做了封装，读者可查看宏 __MMI_TOUCH_PEN_EVENT__ 包含的相关代码。

6.2 铃声播放

在前面我们已经讲解过如何播放一个铃声资源，但是在游戏中我们很少使用铃声资源，而是直接使用铃声的二进制数据。因为播放铃声资源，同一时刻只能播放一个，当另一个铃声资源开始播放后，之前的就被迫停止了。而在游戏中我们需要声音叠加的效果，比如背景音乐播放的过程中，夹杂打击碰撞的声音。那么我们如何得到一个铃声的二进制数据呢？

这里给大家推荐一个文本工具 UltraEdit，在 plutommi\Customer\Audio\PLUTO\MyStudyApp\Games\HitMouse 目录下找到打地鼠游戏的背景音乐文件"wav_mouse_bg.wav"，然后使用 UltraEdit 工具打开，我们看到的那些十六进制数字，就是我们需要的铃声二进制数据，如图 6.2-1 所示，本书使用的 UltraEdit 版本为 18.00.0.1029，读者可以在网上查找更详细的使用方法。

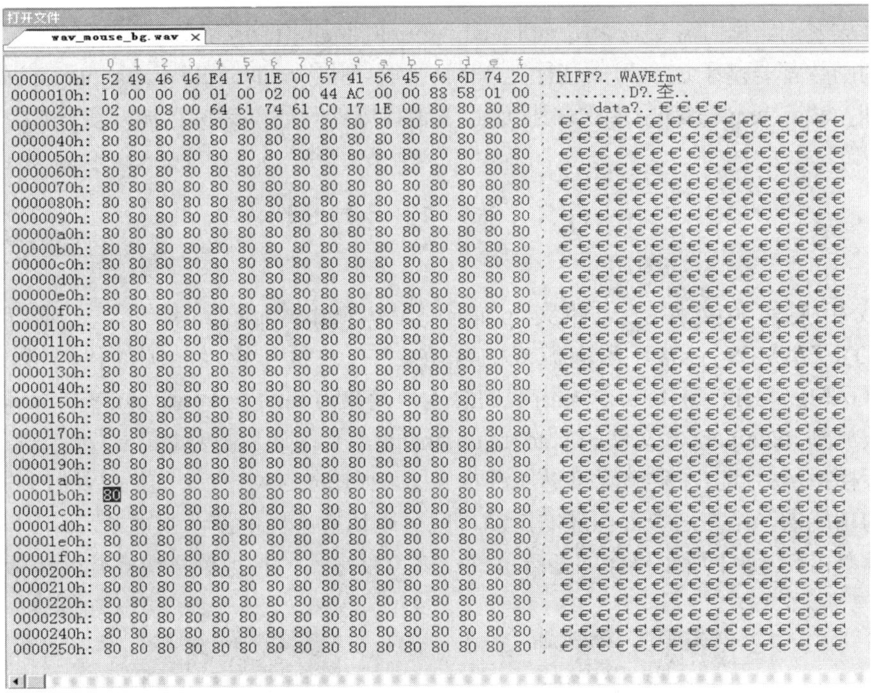

图 6.2-1

按下快捷键 Ctrl+A 全选，依次点击菜单"编辑→十六进制函数→十六进制复制所选查看"，然后新建一个文本，把复制的数据粘贴到新建文本中，再把多余的数据删除，在每个数字前加上"0x"，并用逗号把数字隔开，最终处理的效果如图 6.2-2 所示。

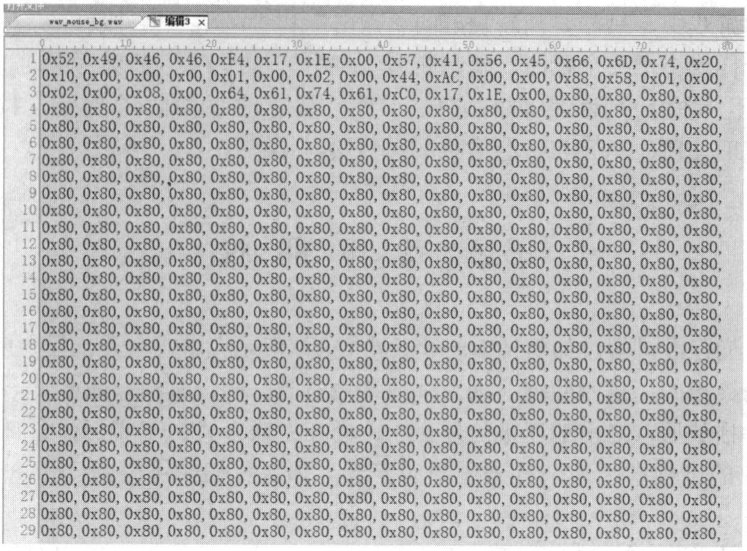

图 6.2-2

接下来只需要把这些数据复制，存储到一个 U8 数组中就可以了。请读者查看源码 plutommi\MyStudyApp\Games\Inc\res_HitMouse.h 文件中的变量 aud_mouse_bg_wav，"打地鼠"游戏的背景音乐就是播放的这个数组。播放涉及的函数可查看 plutommi\Service\MDI\MDISrc\mdi_audio.c 文件中以 "mdi_audio_mma" 开头命名的函数接口。

在播放之前，我们需要先调用 mdi_audio_mma_open_string 函数得到一个铃声句柄，后面所有的函数都是基于这个句柄进行操作的。比如调用 mdi_audio_mma_play 开始播放，调用 mdi_audio_mma_stop 函数停止播放。关于详细的代码，请阅读游戏源码。

6.3 游戏说明

游戏相关的源文件在文件夹 plutommi\MyStudyApp\Games 中，功能宏 _MMI_MYSTUDY_GAMES_ 包含了所有游戏的代码，"打地鼠"游戏相关的代码可以查看宏 __MMI_GAMES_HIT_MOUSE__。游戏的总入口函数为 mmi_custom_games_entry，我们把这个函数放到 mainmenu.c 文件的 goto_main_menu 函数中调用执行。

运行模拟器，点击左软键就可进入游戏界面，如图 6.3-1 所示。如果已经打开了另外两个游戏的功能宏，则在这个界面左右滑动，就可以切换不同的游戏封面了。

鼠标单击图 6.3-1 所示界面的任意位置，就可以进入"打地鼠"游戏开始运行的界面，如图 6.3-2 所示。

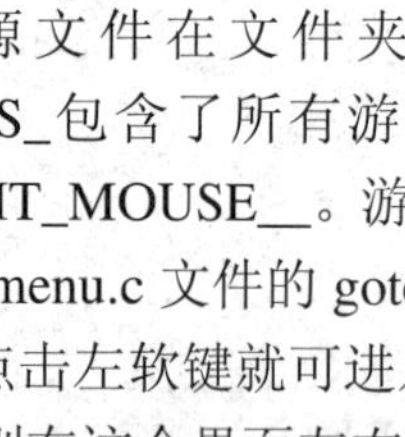

图 6.3-1

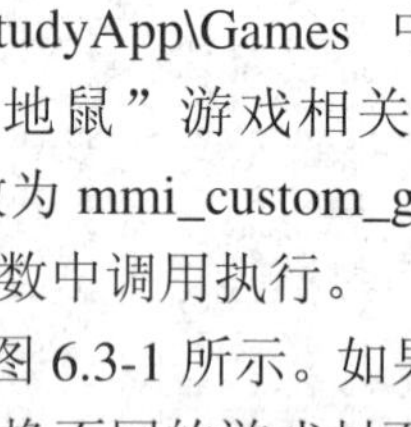

图 6.3-2

各个游戏的入口函数可查看 plutommi\MyStudyApp\Games\Src\ Games.c 文件中的数组变量 g_games_page，其中"打地鼠"游戏的入口函数为"mmi_hit_mouse_entry"，定义在 HitMouse.c 文件中。代码如下(代码清单 6.3-1)：

--代码清单 6.3-1--

```
void mmi_hit_mouse_entry(void)
{
    mmi_frm_scrn_enter(GRP_ID_ROOT,    SCR_HIT_MOUSE_ID,    mmi_hit_mouse_exit,
mmi_hit_mouse_entry, MMI_FRM_FULL_SCRN);

    mmi_hit_mouse_init(g_hitmouse_scene.level);/*初始化关卡*/
    mmi_hit_mouse_redraw_scene();/*绘制场景*/
    mmi_hit_mouse_update_score();/*游戏分数*/
    mmi_hit_mouse_update_time();/*游戏时间*/
    mmi_hit_mouse_update();/*游戏状态更新*/

    if(0 != g_bg_audio_handle)
    {
        mdi_audio_mma_play(g_bg_audio_handle);/*播放游戏背景音乐*/
    }
    wgui_register_pen_up_handler(mmi_hit_mouse_pen_up_hdlr);/*注册触屏弹起事件回调函数*/

    SetRightSoftkeyFunction(mmi_hit_mouse_exit_pop, KEY_EVENT_UP);
                                                        /*退出游戏的提示界面*/
    SetKeyHandler(mmi_hit_mouse_exit_pop, KEY_END, KEY_EVENT_DOWN);
}
```

其中 mmi_hit_mouse_update 函数中有一个定时器，游戏中所有状态的更新都是基于该定时器进行的，包括场景更新、游戏时间更新、地鼠出现或消失的判断。而触屏弹起事件的回调函数 mmi_hit_mouse_pen_up_hdlr 则主要处理玩家点击事件，当点击屏幕时，就会在对应的位置显示锤子动画，只要检测到锤子与地鼠发生碰撞，就给玩家计分并显示地鼠被打中的图片，同时播放打中的声音，然后地鼠钻进洞里，如图 6.3-3 所示。

图 6.3-3

播放锤子动画的函数为 mmi_mouse_hammer，这个函数需要传入老鼠洞的编号，我们规定锤子只能显示在老鼠洞的位置上。代码如下(代码清单 6.3-2)：

--代码清单 6.3-2--

```
void mmi_mouse_hammer(U8 hole)
{
    g_hammer_data.hammer[hole].x = g_mouse_data.mouse[hole].x;
```

```
                g_hammer_data.hammer[hole].y = g_mouse_data.mouse[hole].y;

                gdi_anim_set_draw_before_callback(g_hammer_data.hammer[hole].draw_before_cb);
                gdi_anim_set_last_frame_callback(g_hammer_data.hammer[hole].last_frame_cb);
                gdi_anim_draw_id_once(g_hammer_data.hammer[hole].x,
                g_hammer_data.hammer[hole].y, &IMG_HAMMER_ANIM_ID,
                g_hammer_data.hammer[hole].anim_handle);
                if(0 != g_hit_audio_handle)
                {
                        mdi_audio_mma_play(g_hit_audio_handle);
                }
        }
```

在调用 gdi_anim_draw_id_once 函数播放锤子动画之前，我们调用了 gdi_anim_set_draw_before_callback 函数和 gdi_anim_set_last_frame_callback 函数分别设置动画开始播放前和播放完成后的回调函数，在播放前的回调函数中显示地鼠被打中的效果图片，见函数 mmi_hammer_anim_draw_before_cb_x，比如 mmi_hammer_anim_draw_before_cb_0；而在播放完成后的回调函数中隐藏地鼠并更新游戏分数，见函数 mmi_hammer_anim_finish_cb_x，比如 mmi_hammer_anim_finish_cb_0。游戏中其他与动画相关的状态更新，也是采用此方法实现的。

当游戏结束后，会弹出游戏得分，界面如图 6.3-4 所示。游戏得分的处理函数为 mmi_hit_mouse_update_score，这个函数的内容比较简单，只是显示图片和字符串就可了。

如果玩家按右软键或开关机键退出游戏，则会弹出提示框，让玩家确认是否退出，界面如图 6.3-5 所示。当点击左边的勾时，会退出游戏；而点击右边的叉时，则不退出，继续玩游戏。

图 6.3-4

图 6.3-5

这个界面的入口函数为 mmi_hit_mouse_exit_pop，里面调用了我们在 game.c 文件中封装的一个公共函数接口 mmi_games_pop_screen，代码如下(代码清单 6.3-3)：

---代码清单 6.3-3---
```
        void mmi_games_pop_screen(stu_pop_data *pop_data)
        {
                mmi_confirm_property_struct arg={0x00};
```

```c
        if(NULL == pop_data)
        {
            return;
        }
        g_pop_data.yes_hndl = pop_data->yes_hndl;
        g_pop_data.no_hndl = pop_data->no_hndl;
        g_pop_data.msg_icon = pop_data->msg_icon;

        mmi_confirm_property_init(&arg, CNFM_TYPE_USER_DEFINE);
        arg.parent_id = GRP_ID_ROOT;
        arg.msg_icon = g_pop_data.msg_icon;
        arg.ctype = CNFM_TYPE_YESNO;
        arg.softkey[0].str = (WCHAR *)GetString(STR_GLOBAL_YES);
        arg.softkey[2].str = (WCHAR *)GetString(STR_GLOBAL_NO);
        arg.f_auto_map_empty_softkey = MMI_FALSE;
        arg.callback = (mmi_proc_func)mmi_games_pop_screen_cb;
        mmi_set_pop_bg_img_id(IMG_GAME_POPUP_BG);
        mmi_confirm_display((WCHAR *)L" ", MMI_EVENT_QUERY, &arg);
        wgui_register_pen_up_handler(mmi_games_pop_pen_up_hdlr);
    }
```

这个函数实际上就是调用了系统的提示框函数 mmi_confirm_display，只不过对它做了一些修改，新增了设置背景图片的功能。在 games.c 文件中我们还封装了一些公共函数接口，比如 gui_print_text_ex，这个函数扩展了系统函数 gui_print_text 的功能，把设置字符打印的坐标、颜色、字体都封装到了一个函数中，在调用的时候会方便很多，也能使代码更简洁。

关于打地鼠游戏的完整代码，请读者下载源码阅读。另外，在下载的源码中，笔者还提供了另外两个游戏"翻卡片"和"乐器演奏"，其实现方法都是类似的，希望读者能够从中学到更多的知识。如果有兴趣，可以自己动手写一些简单的小游戏，比如贪吃蛇、推箱子等。

第七章 项目实战

7.1 儿童定位智能电话手表

本节以儿童定位智能电话手表为例，讲解真机环境下 MTK 编程开发方法，带领大家开发一款集打电话、发短信、GPS 定位、联网、屏幕显示等功能于一体的智能电话手表。本节所选用的硬件平台是 MTK6260M，所有实验现象都下载到真机上直接运行。所以，读者在进行本节学习之前，请自备 MTK6260 硬件开发套件，套件获取方式参见第二章。

1. 项目展示

儿童定位智能电话手表除了手表的通用功能外，它还可以获取佩戴者的地理位置，并通过网络上传到服务器，家长则可以通过手机 APP 或微信公众号查看自己的小孩在哪里，家长也可以通过手机拨打手表的电话号码，与小孩进行语音通话。

本节我们讲解基于 MTK6260 平台的儿童定位智能电话手表开发，并通过微信公众号来控制手表完成各项功能，所使用的 MTK 硬件开发套件的外观如图 7.1-1 所示。

为了方便大家学习，我们在"疯壳"微信公众号里申请了一个设备绑定入口，读者可以登录"疯壳"平台，通过该入口将自己的 MTK 硬件开发套件与"疯壳"账号进行绑定，后面所有的实验控制都通过公众号完成。具体的绑定流程如下：

首先，找到该公众号，界面如图 7.1-2 所示。

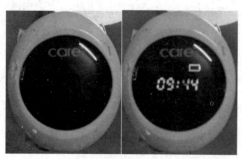

图 7.1-1

图 7.1-2

然后进入该公众号，点击"个人中心"→"我的设备"，界面如图 7.1-3 所示。

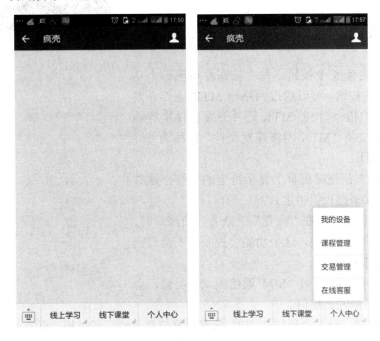

图 7.1-3

(2) 在"我的设备"里面，单击"绑定设备"按钮，进入如图 7.1-4 所示的界面。

这里要求输入一个 32 位的产品 id 码和一个 15 位的 Deviceid 码(即 IMEI 号码)，这两个码都印在 MTK 硬件开发套件的包装袋上，此码非常重要，千万不要遗失。按照提示绑定成功之后，在"我的设备"界面中就会显示绑定的设备列表，如图 7.1-5 所示。

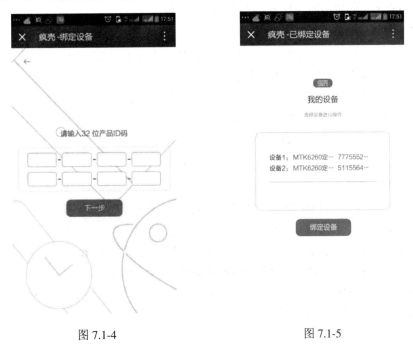

图 7.1-4　　　　　　　　　　图 7.1-5

笔者之前绑定了两个设备，所以这里会列出两个设备的信息。

(3) 点击我们刚才绑定的设备，就会进入如图 7.1-6 所示的控制界面。

这个界面共有 5 个按钮，每个按钮都对应一个功能，点击不同的按钮，可以通过网络给 MTK 硬件开发套件发送不同的指令，控制 MTK 硬件开发套件执行相应的功能。(下文将"MTK 硬件开发套件"简称为"终端")，具体如下：

① 屏幕测试：终端屏幕上显示特定的字符，这些字符可以在终端通过代码指定。

② SIM 通信测试：在"疯壳"公众号注册账户时，要求填写自己的手机号码，这个功能会控制手表给自己的手机号码拨打电话。

图 7.1-6

③ SIM 短信测试：同"SIM 通信测试"类似，会给自己的手机号码发送短信。

④ GPS 测试：设备会先打开 GPS 搜索位置信息，然后通过短信的方式，每隔一分钟发送一次 GPS 经纬度数据，最多发送 10 次，如果搜索到了经纬度或者超过 10 分钟还没搜索到经纬度则关闭 GPS。在测试此功能时，一定要把手表放在室外空旷处，GPS 在室内是无法定位的。

⑤ 手表对时：同步服务器时间。

以上几个功能，在测试的时候，一定要先测试 SIM 通信或 SIM 短信功能，再测试 GPS 功能，否则 GPS 发送短信时无法获取短信接收号码。

2．项目需求

从项目展示的内容中可以总结出该项目在软件上主要有以下几个需求：
(1) 屏幕内容显示：显示时间以及其他内容。
(2) 网络通信功能：与服务器进行交互，响应微信公众号中的指令。
(3) 通话功能：手表通过 SIM 卡可以拨打电话。
(4) 短信功能：手表通过 SIM 卡可以发送短信。
(5) GPS 定位功能：通过 GPS 获取当前位置信息。

3．网络通信协议

网络通信协议采用 HTTP 协议，服务器接收到请求的返回值都是 JSON 格式，设备和服务器采用常轮询的方式模拟长连接。手表每隔两秒给服务器发送心跳包，如果服务器上有 APP 端推送过来的指令，则返回手表端，如果没有则应答心跳指令。

服务器地址：www.fengke.club/GeekMartMobile/device/req.json?
服务器端口号：80
手表端请求协议格式见表 7.1。

表 7.1　手表端请求协议格式

请求值			返回值		
Cmd	ReqVal	deviceid	Cmd	ConfirmCode	resqVal
"heart"	030	设备 IMEI 号码	"heart"	031	-1

指令示例："cmd=heart&reqVal=030&deviceId=123456789000000"

服务器端返回的指令协议格式见表 7.2。

表 7.2　服务器端返回指令协议格式

请求值			返回值
Cmd	ConfirmCode	resqVal	
"test"	230(屏幕测试)	-1	无
"test"	231(SIM 通信测试)	-1	
"test"	232(SIM 短信测试)	-1	
"test"	233(GPS 测试)	-1	
"test"	234(手表对时)	年月日时分	

指令示例：

屏幕测试：{"cmd":"test", "confirmCode":"230", "respVal":"-1"}

SIM 通信测试：{"cmd":"test", "confirmCode":"231", "respVal":"10086"}

SIM 短信测试：{"cmd":"test", "confirmCode":"232", "respVal":"10086"}

GPS 测试：{"cmd":"test", "confirmCode":"233", "respVal":"-1"}

手表对时：{"cmd":"test", "confirmCode":"234", "respVal":"201712301200"}

4. 代码实现

在前面的学习过程中，我们添加了一个功能模块——MystudyApp，它不仅实现了网络通信、打电话、发短信、定时器、GPS 定位等功能，而且还移植了 JSON 协议。接下来我们就在前面实现的功能基础上完成儿童定位智能手表上的 Demo 功能。

1) 项目配置

每个项目都应该有一个项目宏，项目宏的作用是包含所有项目差异化的代码，包括驱动代码和 MMI 代码，其使用优先级应低于功能宏。我们通常在项目配置文件 make\UMEOX60M_11B_GPRS.mak 中定义项目名称，参照宏 MYSTUDY_APPLICATION 新增宏 PROJECT_NAME，然后指定项目名称为 CHILDREN_WATCH，代码如下(代码清单 7.1-1)：

--代码清单 7.1-1--

/*省略系统默认代码*/

#--##
#---------- mystudy add begin
MYSTUDY_APPLICATION = TRUE

```
PROJECT_NAME = CHILDREN_WATCH ##  项目名称


#---------- custom add end
#---------------------------------------------------##
```

/*省略系统默认代码*/

然后在 make\Option.mak 文件中根据项目名称定义项目宏 _PROJ_CHILDREN_WATCH_。代码如代码清单 7.1-2 所示，其中代码表达式 $(strip $(PROJECT_NAME)) 的作用是取 PROJECT_NAME 宏定义的值，以字符串的形式表示。

---代码清单 7.1-2---

/*省略系统默认代码*/

```
#------------------__MYSTUDY_APPLICATION__----------------##
#---------- mystudy add begin


# **********************************************************
# 功能宏
# **********************************************************
ifdef MYSTUDY_APPLICATION
    ifeq ($(strip $(MYSTUDY_APPLICATION)), TRUE)
        COMPLIST         += mystudyapp
        CUS_REL_BASE_COMP += make\mystudyapp
        COM_DEFS         += __MYSTUDY_APPLICATION__
    endif
endif

ifneq (, $(strip $(PROJECT_NAME)))
    COM_DEFS += __PROJ_$(strip $(PROJECT_NAME))__    ## 定义项目宏
endif


#---------- mystudy add end
#---------------------------------------------------##
```

该项目屏幕尺寸为 64×48，而 MTK 平台默认支持的屏幕最小分辨率为 96×64。通常这么小的屏幕尺寸显示的内容非常有限，不会在界面上做太复杂的功能，所以我们不需要定制一个 64×48 的界面兼容所有的 UI 显示。在项目配置文件 make\UMEOX60M_11B_GPRS.mak 中设置 MAIN_LCD_SIZE = 96×64，然后在 custom\common\hal_public\lcd_sw_inc.h 文件中设置 LCD_WIDTH 和 LCD_HEIGHT 宏定义值分别为 64、48。代码如下(代码清单 7.1-3)：

--代码清单 7.1-3--
```
#if defined(_MMI_MAINLCD_96X64_)
#if (defined(_PROJ_CHILDREN_WATCH__)||defined(_MYSTUDY_APPLICATION_))
    #define LCD_WIDTH          (64)          /*屏幕宽度*/
    #define LCD_HEIGHT         (48)          /*屏幕高度*/
#else
    #define LCD_WIDTH          (96)
    #define LCD_HEIGHT         (64)
#endif
#elif defined(_MMI_MAINLCD_128X128_)
```

这样就把屏幕的分辨率改成了 64×48，但是在编译过程中肯定还有其他地方需要修改，只不过比起重新新建一个屏幕尺寸，改动的地方会少很多。读者可在代码中全局搜索项目宏_PROJ_CHILDREN_WATCH_，查看修改的地方。

2) 屏幕显示

在前面讲解界面显示的章节中，我们在屏幕上打印字符、显示图片等都可以调用 MTK 源码提供的函数接口，这种绘制界面的方式是 MTK 平台上比较常用的。但在实际做项目的过程中，我们总会碰到各种各样的特例。比如我们这个儿童手表，其屏幕尺寸只有 64×48，因其显示内容比较简单，而且比较少，所以我们的屏幕没有兼容 MTK 上的界面绘制接口，而是直接使用取模工具把需要显示的内容转换成二进制点阵数组，然后直接调用屏幕驱动的描点函数接口绘制到屏幕上，这样做的好处就是屏幕刷新速度很快。由于本内容有点偏门，而且跟屏幕驱动相关，属于驱动工程师的工作范畴，这里就不深入探究，只要了解怎么用就可以了。下面先给大家介绍一个取字模的工具，读者可自己在网上下载，打开后界面显示如图 7.1-7 所示，具体操作步骤如下。

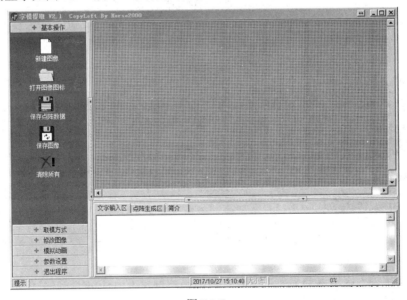

图 7.1-7

(1) 在参数设置中单击"文字输入区字体选择",设置字体为"宋体""常规""小四",如图 7.1-8 所示。

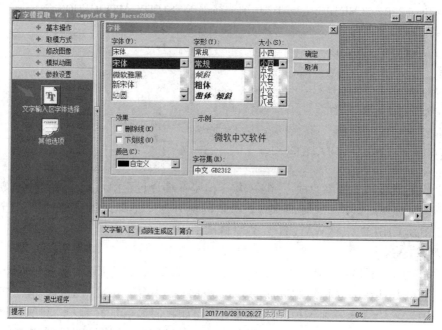

图 7.1-8

(2) 单击参数设置中的"其他选项",设置详情如图 7.1-9 所示。

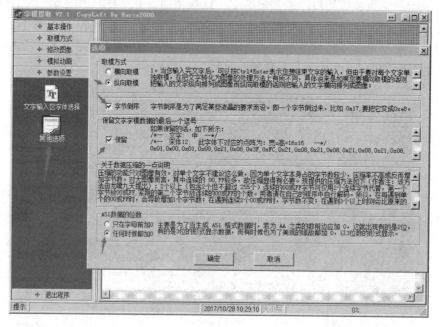

图 7.1-9

(3) 在文字输入区输入要显示的字符,最后按快捷键 Ctrl+Enter,就会生成对应的点阵图像,比如输入"屏幕"两个汉字,如图 7.1-10 所示。

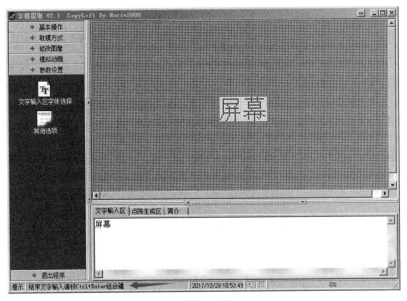

图 7.1-10

(4) 单击展开"取模方式"菜单,再点击"C51 格式",就会看到如图 7.1-11 所示的界面。在点阵生成区里面,就是汉字"屏"和"幕"的点阵数据。

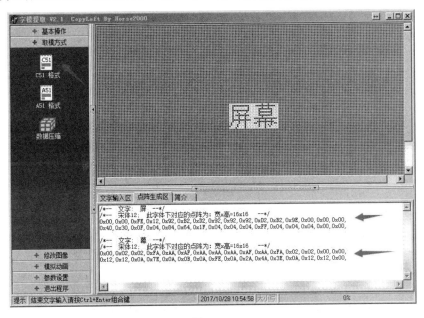

图 7.1-11

接下来,我们把生成的点阵数据复制粘贴到代码中,定义一个 unsigned char 类型的数组保存它,代码如下(代码清单 7.1-4):

--代码清单 7.1-4--

const unsigned char str_ping[] ={/*-- 文字: 屏 --*/

/*-- 宋体 12; 此字体下对应的点阵为:宽 x 高=16x16   --*/

0x00, 0x00, 0xFE, 0x12, 0x92, 0xB2, 0xD2, 0x92, 0x92, 0x92, 0xD2, 0xB2, 0x9E, 0x00, 0x00, 0x00,

```
    0x40, 0x30, 0x0F, 0x04, 0x84, 0x64, 0x1F, 0x04, 0x04, 0x04, 0xFF, 0x04, 0x04, 0x04, 0x00, 0x00,
    0x00, 0x00, 0x00, 0x00, 0x00, 0x00, 0x00, 0x00, 0x00, 0x00, 0x00, 0x00, 0x00, 0x00, 0x00, 0x00,
};
const unsigned char str_mu[] ={/*--  文字:   幕   --*/
        /*--  宋体 12;  此字体下对应的点阵为:宽 x 高=16x16  --*/
    0x00, 0x02, 0x02, 0xFA, 0xAA, 0xAF, 0xAA, 0xAA, 0xAA, 0xAF, 0xAA, 0xFA, 0x02, 0x02, 0x00,0x00,
    0x12, 0x12, 0x0A, 0x7E, 0x0A, 0x0B, 0x0A, 0xFE, 0x0A, 0x2A, 0x4A, 0x3E, 0x0A, 0x12, 0x12, 0x00,
    0x00, 0x00, 0x00, 0x00, 0x00, 0x00, 0x00, 0x00, 0x00, 0x00, 0x00, 0x00, 0x00, 0x00, 0x00, 0x00,
};
```

在上面的数组中,除了字模点阵数据外,还添加了 16 个字节的 0x00。因为我们的屏幕高度只有 48 个像素,基本上只能显示一行字符,而取出的字模数据总共是 32 个字节,所以我们在数组中填充 16 个 0x00,补齐 48 个字节。以上两个数组,可以通过屏幕驱动的接口把它们显示出来。我们的屏幕上每次只显示两个字符,笔者已经封装了一个函数接口,代码如下(代码清单 7.1-5):

--代码清单 7.1-5--
```
void mmi_show_two_words(const unsigned char *word1, const unsigned char *word2)
{
    unsigned char x=0, y=0;
    unsigned int  j=0;

    if(NULL==word1 || NULL==word2)
    {
        return;
    }
    mmi_clean_screen();    /*清除屏幕*/

    /*显示第一个字符*/
    SSD1306_SendCmd(0x08);         /* set  lower  column  address    */
    SSD1306_SendCmd(0x12);         /* set  higter column  address    */
    for(y=1, j=0;y<4;y++)
    {
        SSD1306_SendCmd(0xb0+y); /* set  page  address    */
        SSD1306_SendCmd(0x08);     /* set  lower  column  address    */
        SSD1306_SendCmd(0x12);     /* set  higter column  address    */
        for(x=0;x<16;x++)
        {
            SSD1306_SendData(word1[j++]);
        }
    }
```

```
           /*显示第二个字符*/
           SSD1306_SendCmd(0x08);         /*  set  lower  column  address  */
           SSD1306_SendCmd(0x14);         /*  set  higter column  address  */
           for(y=1, j=0;y<4;y++)
           {
                SSD1306_SendCmd(0xb0+y);  /*  set  page   address          */
                SSD1306_SendCmd(0x08);    /*  set  lower  column  address  */
                SSD1306_SendCmd(0x14);    /*  set  higter column  address  */
                for(x=0;x<16;x++)
                {
                     SSD1306_SendData(word2[j++]);
                }
           }
     }
```

3) 框架搭建

我们依旧使用 MystudyApp 的框架,这个框架有一个专业术语叫"MVC 框架",即模型(Model)－视图(View)－控制器(Controller),其中定位模块、Socket 网络通信模块的代码我们已经写好了,可以直接拿来使用。在 MyStudyAppMain.c 文件中编写逻辑控制代码,在 MyStudyAppMain 目录下对应的文件夹中新建 res_screendata.c 和 res_screendata.h 文件用于管理屏幕上待显示内容的点阵数据资源,同样在 MyStudyAppMain 目录下对应的文件夹中新建 ScreenDisplay.c 和 ScreenDisplay.h 文件,用于显示屏幕上的内容。

关于屏幕绘制,这里用到了一个全局的函数指针 mmi_screen_update_redraw,在绘制界面的时候,我们通过改变它的值来实现界面的切换,其默认值为 showpic_idle_timeall,这个函数就是开机后显示时间界面的函数。我们在 plutommi\mmi\Idle\IdleSrc\idleclassic.c 文件的 mmi_idle_classic_on_init 函数中把 idle 界面的绘制函数替换为 mmi_digitclock_draw 函数,而 mmi_digitclock_draw 函数中调用了 mmi_screen_update_redraw 函数指针,执行默认的函数 showpic_idle_timeall。

关于与服务器的网络交互功能,其入口函数为 MyStudyAppMain.c 文件中的 mmi_mystudy_app_init 函数,这个函数一开机就会执行,我们在里面启动了一个定时器,然后在定时器回调函数中不断地再次启动,每隔两秒执行发送一次心跳包,代码如下(代码清单 7.1-6):

```
-------------------------------------代码清单 7.1-6-------------------------------------
static void mmi_net_cmd_timer_handle(void)
{
     g_netcmd_data[0].handle(CLIENT_MSG, NULL);/*发送心跳包*/

     StartTimer(NET_CMD_HANDLE_TIMER, 2*1000, mmi_net_cmd_timer_handle);
     /*每隔两秒执行一次*/
```

```
}

/*程序入口函数*/
void mmi_mystudy_app_init(void)
{
    /*检测 SIM 卡*/
    if(MMI_FALSE==srv_sim_ctrl_is_inserted(MMI_SIM1)||
MMI_FALSE==srv_sim_ctrl_is_available(MMI_SIM1))
    {
        kal_prompt_trace(MOD_XDM, "--%d(sim invalid:%d, %d)--%s--",
        _LINE_, srv_sim_ctrl_is_inserted(MMI_SIM1),
        srv_sim_ctrl_is_available(MMI_SIM1), _FILE_);
        StartTimer(NET_CMD_HANDLE_TIMER, 5*1000, mmi_mystudy_app_init);
        return;
    }

    socket_socket_init();/*socket 初始化，通常也可以放在 mmi_bootup_notify_completed 函数中调用*/
    mmi_socket_set_callback(mmi_mystudy_socket_cb);    /*设置 socket 回调函数*/
    StartTimer(NET_CMD_HANDLE_TIMER, 10*1000, mmi_net_cmd_timer_handle);
                    /*开机 10 秒后开始联网，执行网络通信*/
}
#endif
```

关于完整的详细代码，请读者上网下载本书附带的源码。

5．发布版本

在项目开发过程中，每一次需求完成都需要发布软件版本，在 make 目录下的项目版本控制文件中更新版本号。本项目中打开 make\Verno_UMEOX60M_11B.bld 文件，修改版本号为"Children_Wacth_V0.01"，代码如下(代码清单 7.1-7)：

```
-----------------------------------代码清单 7.1-7-----------------------------------
VERNO    =    Children_Wacth_V0.01
HW_VER   =    UMEOX60M_11B_HW
BUILD    =    BUILD_NO
BRANCH   =    11BW1308MP
----------------------------------------------------------------------------
```

其中"Children_Wacth"为版本名称，通常与项目名称相关，"V0.01"为版本编号，这个编号是可以递增的。需要注意的是，这里所说的项目名称并不是代码中所说的项目名称，而是产品的项目名称，比如代码中的项目名称为"UMEOX60M_11B"，而产品的项目名称为"Children_Wacth"。版本命名并没有严格的规定，通常由客户指定。

修改了版本号，需要执行 make new 指令重新编译代码。编译完成后我们再把生成的软件版本打包，软件版本相关的文件有两部分：一部分是 Catcher 中打 log 需要用到的数据库文件，这部分文件只有一个，为 tst\database_classb 目录下的 BPLGUInfoCustomAppSrcP_MT6260_S00_CHILDREN_WACTH_V0_01 文件；另一部分是 Flash_Tool 工具中下载需要用到的文件，在 build 目录中对应的项目名称文件夹(UMEOX60M_11B\UMEOX60M_11B_PCB01_gprs_MT6260_S00.CHILDREN_WACTH_V0_01.bin)中，如果不确定是哪些文件，有一个很简单的办法，就是打开 Flash_Tool 把下载的文件全部加载进去，就能很清楚地看到哪些文件是下载需要的，如图 7.1-12 所示。

图 7.1-12

这里需要的文件有 5 个，分别为：ROM、UMEOX60M_11B_BOOTLOADER_V005_MT6260_CHILDREN_WACTH_V0_01.bin、VIVA、UMEOX60M_11B_BB.cfg、EXT_BOOTLOADER。加上 Catcher 中用到那个文件，在版本发布中需包含 6 个文件，如图 7.1-13 所示。需要注意的是，不同的 MTK 平台下，版本相关的文件是不一样的。

图 7.1-13

6. 写 IMEI 号码

IMEI(International Mobile Equipment Identity)是国际移动设备身份码的缩写，通常在购买任意移动设备(比如手机、智能手表)时，这个号码都会印在包装盒或设备上的某个地方。

这个编码可以通过工具写进设备中，然后可以在代码中通过相关函数接口读取出来。在服务器端通常使用 IMEI 号码作为设备的唯一标志。读者在购买到手表的时候，一定要备份好这个号码，否则无法访问服务器。

写 IMEI 号码的工具叫做"SN_Writer"，在下载的工具中，压缩包名称为"SN_Writer_Tool_V1.1632.00.rar"，这个工具不需要安装，使用步骤如下：

解压后双击"SN Writer.exe"就可以打开，界面显示如图 7.1-14 所示。

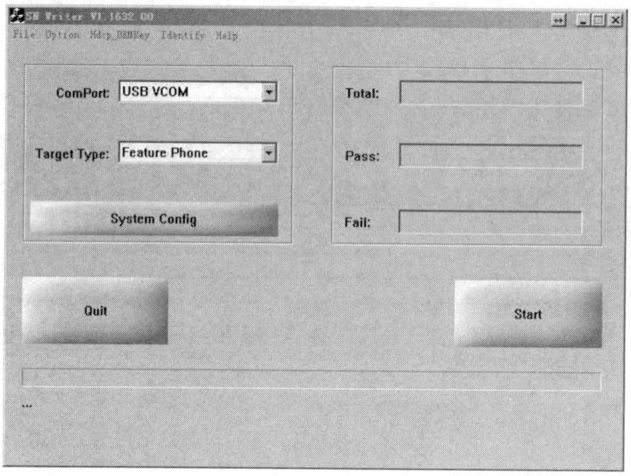

图 7.1-14

（2）单击"System Config"按钮，在弹出的设置界面中，按图 7.1-15 所示的界面进行设置。其中单击"MD1_DB"按钮，在弹出的文件选择框中，需要选择 Catcher 中加载的那个数据库文件：BPLGUInfoCustomAppSrcP_MT6260_S00_CHILDREN_WACTH_V0_01。设置完成后，单击"Save"按钮。

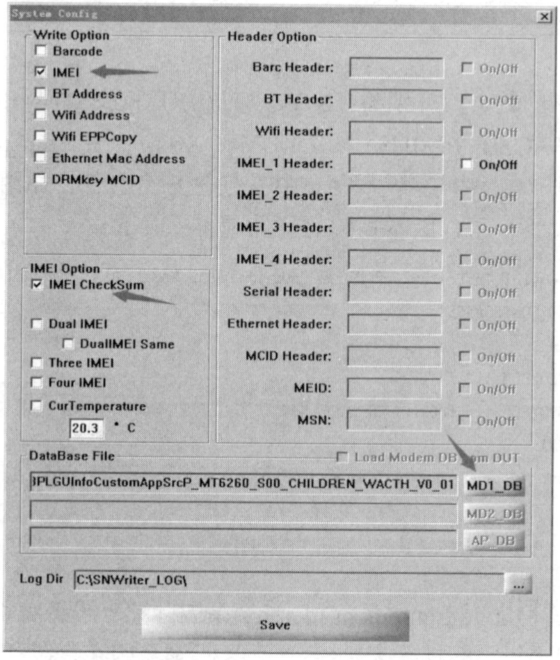

图 7.1-15

(3) 单击"Target Type"下拉选择框,选择"Feature Phone",再单击"Start"按钮,在弹出的对话框中输入要写入的 IMEI 号码,这个 IMEI 号必须是自己购买手表时印在包装盒上的 IMEI 号码,不能随便输入,如图 7.1-16 所示。

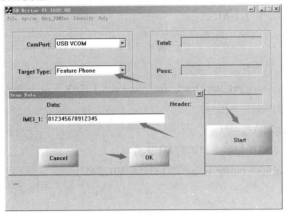

图 7.1-16

(4) 单击"OK"按钮,会出现如图 7.1-17 所示的界面,此时先把设备关机,然后用 USB 线连接电脑,与烧录版本时的步骤是一样的。

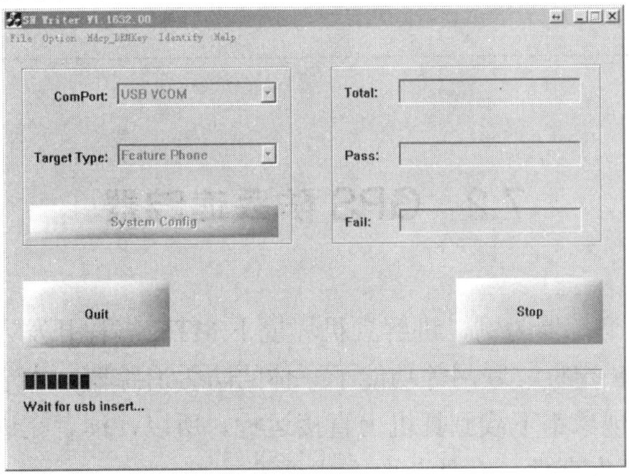

图 7.1-17

写成功之后,就会出现如图 7.1-18 所示的界面。

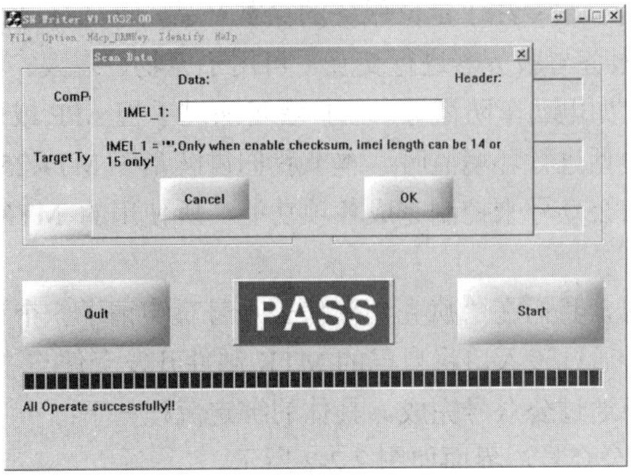

图 7.1-18

IMEI 号通常只要写入一次就可以了,但是在下载版本的时候,如果选择了格式化(Format FAT)下载,则需要重新写入。如果在模拟器上调试网络部分的代码,则可以在获取 IMEI 的函数接口 srv_imei_get_imei 中(plutommi\Service\ImeiSrv\ImeiSrv.c),手动输入 IMEI 号码,代码如下(代码清单 7.1-8):

```
-----------------------------------------代码清单 7.1-8-----------------------------------------
/*省略部分默认代码*/
MMI_BOOL srv_imei_get_imei(mmi_sim_enum sim, CHAR *out_buffer, U32 buffer_size)
{
    /*省略部分默认代码*/
#ifdef WIN32/*设置模拟器中调试的 IMEI 号码*/
    strcpy((char*)out_buffer, (char*)"012345678912345");
#else
    app_strlcpy(out_buffer, cntx->imei, buffer_size);
#endif
    return (out_buffer[0] == '\0' ? MMI_FALSE : MMI_TRUE);
}
-----------------------------------------------------------------------------------------------
```

7.2 GPS 防丢追踪器

本节以 GPS 防丢追踪器为例,讲解真机环境下 MTK 编程开发方法,带领大家开发一款集电话、短信、GPS 定位、联网等功能于一体的防丢追踪器。本节所选用的硬件平台是 MTK2502,所有实验现象都下载到真机上直接运行。所以,读者在进行本节学习之前,请自备 MTK2502 硬件开发套件,套件获取方式见第二章。

1. 项目展示

GPS 防丢追踪器是一款专注于定位服务的穿戴产品,它具有电话、短信、定位等功能,主要通过 SIM 卡流量服务与服务器进行交互,可用于宠物、老人、小孩身上,也可以扩展到一些行业应用中,比如电动车防盗等。监护者可通过手机 APP 或微信公众号实时查看设备的位置,可以拨打电话进行环境监听。本节我们讲解基于 MTK2502 平台的 GPS 防丢追踪器开发,并通过微信公众号来控制完成各项功能,所使用的 MTK 硬件开发套件设备外观如图 7.2-1 所示。

为了方便大家学习,我们在"疯壳"微信公众号里申请了一个设备绑定入口,读者可以登录"疯壳"平台,通过该入口将自己的 MTK 硬件开发套件与"疯壳"账号进行绑定,后面所有的实验控制都通过公众号完成。具体的绑定流程如下:

(1) 首先,找到该公众号,界面如图 7.2-2 所示。

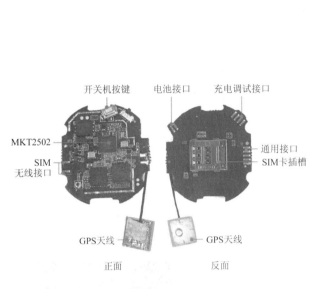

图 7.2-1　　　　　　　　　　　　　　图 7.2-2

然后进入"疯壳"微信公众号,点击"个人中心"→"我的设备",界面如图 7.2-3 所示。

图 7.2-3

(2) 在"我的设备"里面,单击"绑定设备"按钮,进入如图 7.2-4 所示的界面。

这里要求输入一个 32 位的产品 ID 码和一个 15 位的 Deviceid 码(即 IMEI 号码),这两个码都印在 MTK 硬件开发套件的包装袋上,此码非常重要,千万不要遗失。按照提示绑定成功之后,在"我的设备"界面中就会显示绑定的设备列表,如图 7.2-5 所示。

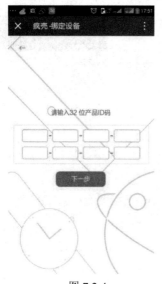

图 7.2-4　　　　　　　　　　图 7.2-5

笔者之前绑定了两个设备，所以这里会列出两个设备的信息。

（3）点击我们要调试的设备，就会进入如图 7.2-6 所示的界面。

这个界面共有 5 个按钮，每个按钮都对应一个功能，点击不同的按钮，可以通过网络给终端发送不同的指令，控制终端执行相应的功能。因为该项目不带屏幕，关于屏幕显示的内容我们都无法看到效果，所以"屏幕测试"功能忽略不计。具体功能如下：

① SIM 通信测试：在"疯壳"公众号注册账户时，要求填写自己的手机号码，这个功能会控制手表给自己的手机号码拨打电话。

② SIM 短信测试：同"SIM 通信测试"类似，会给自己的手机号码发送短信。

③ GPS 测试：设备会先打开 GPS 搜索位置信息，然后通过短信的方式,每隔一分钟发送一次 GPS 经纬度数据，

图 7.2-6

最多发送 10 次，如果搜索到了经纬度或者超过 10 分钟还没搜索到经纬度则关闭 GPS。在测试此功能时，一定要把手表放在室外空旷处，GPS 在室内是无法定位的。

④ 手表对时：同步服务器时间。因为该设备不会显示时间，所以在测试 GPS 和短信的时候，我们把系统当前时间通过短信一起发送。

以上几个功能，在测试的时候，一定要先测试 SIM 通信或 SIM 短信功能，再测试 GPS 功能，否则 GPS 发送短信时，无法获取短信接收号码。时间同步功能也可以先于 GPS 测试和短信测试。

2．项目需求

从项目展示的内容中可以总结出该项目在软件上主要有以下几个需求：

(1) 网络通信功能：与服务器进行交互，响应微信公众号中的指令。
(2) 通话功能：手表通过 SIM 卡可以拨打电话。
(3) 短信功能：手表通过 SIM 卡可以发送短信。
(4) GPS 定位功能：通过 GPS 获取当前位置信息。

3．网络通信协议

网络通信协议采用 HTPP 协议，服务器接收到请求的返回值都是 JSON 格式，设备和服务器采用长轮询的方式模拟长连接。手表每隔两秒给服务器发送心跳包，如果服务器上有 APP 端推送过来的指令，则返回手表端，如果没有则应答心跳指令。

服务器地址：www.fengke.club/GeekMartMobile/device/req.json?

服务器端口号：80

手表端请求协议格式见表 7.3。

表 7.3　手表端请求协议格式

请求值			返回值		
Cmd	ReqVal	deviceid	Cmd	ConfirmCode	resqVal
"heart"	030	设备 IMEI 号码	"heart"	031	−1

指令示例："cmd=heart&reqVal=030&deviceId=123456789000000"

服务器端返回的指令协议格式见表 7.4。

表 7.4　服务器端返回指令协议格式

请求值			返回值
Cmd	ConfirmCode	resqVal	
"test"	231(SIM 通话)	−1	无
"test"	232(SIM 短信)	−1	
"test"	233(GPS 测试)	−1	
"test"	234(手表校时)	年月日时分	

指令示例：

SIM 通话：{"cmd":"test", "confirmCode":"231", "respVal":"10086"}

SIM 短信：{"cmd":"test", "confirmCode":"232", "respVal":"10086"}

GPS 测试：{"cmd":"test", "confirmCode":"233", "respVal":"-1"}

时间同步：{"cmd":"test", "confirmCode":"234", "respVal":"201712301200"}

4．代码实现

在前面的学习过程中，我们添加了一个功能模块——MystudyApp，它不仅实现了网络通信、打电话、发短信、定时器、GPS 定位等功能，而且还移植了 JSON 协议。接下来我们就在前面实现的功能基础上完成防丢追踪器上的 demo 功能。

1) 项目配置

每个项目都应该有一个项目宏，项目宏的作用是包含所有项目差异化的代码，包括驱动代码和 MMI 代码，其使用优先级应低于功能宏。我们通常在项目配置文件 make\HEXING02A_WT_11C_GPRS.mak 中定义项目名称,参照宏 MYSTUDY_APPLICATION 新增宏 PROJECT_NAME，然后指定项目代号为 FK001，代码如代码清单 7.2-1 所示。

--代码清单 7.2-1--
/*省略系统默认代码*/

#--##
#---------- mystudy add begin

MYSTUDY_APPLICATION = TRUE

PROJECT_NAME = FK001 ## 项目名称

#---------- custom add end
#--##

/*省略系统默认代码*/

然后在 make\Option.mak 文件中根据项目名称定义项目宏_PROJ_FK001__。代码如代码清单 7.2-2 所示，其中代码表达式$(strip $(PROJECT_NAME))的作用是取 PROJECT_NAME 宏定义的值，以字符串的形式表示。

--代码清单 7.2-2--
/*省略系统默认代码*/
#------------------_MYSTUDY_APPLICATION_----------------##
#---------- mystudy add begin

**
功能宏
**
ifdef MYSTUDY_APPLICATION
    ifeq ($(strip $(MYSTUDY_APPLICATION)), TRUE)
        COMPLIST   += mystudyapp
        CUS_REL_BASE_COMP += make\mystudyapp
        COM_DEFS   += _MYSTUDY_APPLICATION_
    endif
endif

ifneq (, $(strip $(PROJECT_NAME)))
    COM_DEFS += _PROJ_$(strip $(PROJECT_NAME))_   ## 定义项目宏
endif
#---------- mystudy add end
#--##

因为该项目不带屏幕,所以屏幕尺寸设置为多少都没有关系。关于项目功能的实现源码,读者可在代码中全局搜索项目宏_PROJ_FK001_,查看修改的地方。

2) 框架搭建

我们依旧使用 MystudyApp 的框架,这个框架有一个专业术语叫"MVC 框架",即模型(Model)-视图(View)-控制器(Controller),其中定位模块、Socket 网络通信模块的代码我们已经写好了,可以直接拿来使用。在 MyStudyAppMain.c 文件中编写逻辑控制代码。

关于与服务器的网络交互功能,其入口函数为 MyStudyAppMain.c 文件中的 mmi_mystudy_app_init 函数,这个函数一开机就会执行,我们在里面启动了一个定时器,然后在定时器回调函数中不断地再次启动,每隔两秒执行发送一次心跳包,代码如代码清单 7.2-3 所示。

---代码清单 7.2-3---

```
static void mmi_net_cmd_timer_handle(void)
{
    g_netcmd_data[0].handle(CLIENT_MSG, NULL);/*发送心跳包*/
    StartTimer(NET_CMD_HANDLE_TIMER, 2*1000, mmi_net_cmd_timer_handle);
        /*每隔两秒执行一次*/
}


/*程序入口函数*/
void mmi_mystudy_app_init(void)
{
    /*检测 SIM 卡*/
    if(MMI_FALSE==srv_sim_ctrl_is_inserted(MMI_SIM1)
     || MMI_FALSE==srv_sim_ctrl_is_available(MMI_SIM1))
    {
        kal_prompt_trace(MOD_XDM, "--%d(sim check:%d, %d)--%s--",
            _LINE_, srv_sim_ctrl_is_inserted(MMI_SIM1),
            srv_sim_ctrl_is_available(MMI_SIM1), _FILE_);
        StartTimer(NET_CMD_HANDLE_TIMER, 5*1000, mmi_mystudy_app_init);
        return;
    }

    socket_socket_init();/*socket 初始化,通常也可以放在 mmi_bootup_notify_completed 函数中调用*/
    mmi_socket_set_callback(mmi_mystudy_socket_cb);   /*设置 socket 回调函数*/
    StartTimer(NET_CMD_HANDLE_TIMER, 10*1000, mmi_net_cmd_timer_handle);
            /*开机 10 秒后开始联网,执行网络通信*/
}
#endif
```

关于完整的详细代码，请读者上网下载本书附带的源码。

5．发布版本

在项目开发过程中，每一次需求完成，都需要发布软件版本，在 make 目录下的项目版本控制文件中更新版本号。本项目中打开 make\Verno_HEXING02A_WT_11C.bld 文件，修改版本号为"FK001_V0.01"，代码如代码清单 7.2-4 所示。

---------------------------------------代码清单 7.2-4---------------------------------------

```
VERNO    =    FK001_V0.01
HW_VER   =    HEXING02A_WT_11C_HW
BUILD    =    1
BRANCH   =    11CW1418SP4
```

其中"FK001"为版本名称，或者称为项目代号，通常与项目名称相关，可以按照任意规则定义，"V0.01"为版本编号，这个编号是可以递增的。需要注意的是，这里所说的项目名称并不是代码中所说的项目名称，而是产品的项目名称，比如代码中的项目名称为"HEXING02A_WT_11C"，而产品的项目名称为"FK001"。版本命名并没有严格的规定，通常由客户指定。

修改了版本号，需要执行 make new 指令重新编译代码。编译完成后我们再把生成的软件版本打包。软件版本相关的文件有两部分：一部分是 catcher 中打 log 需要用到的数据库文件，这部分文件只有一个，为 tst\database_classb 目录下的 BPLGUInfoCustomAppSrcP_MT2502_S00_FK001_V0_01 文件；另一部分是 Flash_Tool 工具中下载需要用到的文件，在 build 目录中对应的项目名称文件夹 (HEXING02A_WT_11C\HEXING02A_WT_11C_PCB01_gprs_MT2502_S00.FK001_V0_01.bin)中，如果不确定是哪些文件，有一个很简单的办法，就是打开 Flash_Tool 把下载的文件全部加载进去，就能很清楚地看到哪些文件是下载需要的，如图 7.2-7 所示。

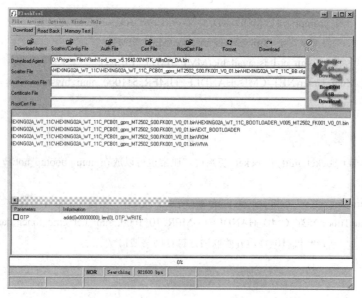

图 7.2-7

这里需要的文件有 5 个，如图 7.2-8 所示。需要注意的是，不同的 MTK 平台下，版本相关的文件名称和个数都有可能是不一样的。

```
EXT_BOOTLOADER
HEXING02A_WT_11C_BB.cfg
HEXING02A_WT_11C_BOOTLOADER_V005_MT2502_FK001_V0_01.bin
ROM
VIVA
```

图 7.2-8

6. 写 IMEI 号码

IMEI(International Mobile Equipment Identity)是国际移动设备身份码的缩写，通常在购买任意移动设备(比如手机、智能手表)时，这个号码都会印在包装盒或设备上的某个地方。这个编码可以通过工具写进设备中，然后可以在代码中通过相关函数接口读取出来。在服务器端通常使用 IMEI 号码作为设备的唯一标志。读者在购买到手表的时候，一定要备份好这个号码，否则无法访问服务器。

写 IMEI 号码的工具叫做"SN_Writer"，在下载的工具中，压缩包名称为"SN_Writer_Tool_V1.1632.00.rar"，这个工具不需要安装，使用步骤如下：

(1) 解压后双击"SN Writer.exe"就可以打开，界面显示如图 7.2-9 所示。

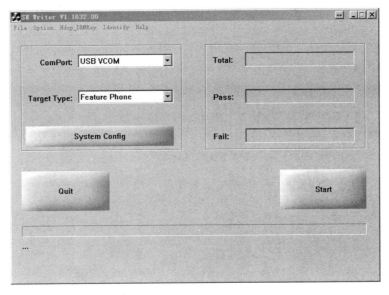

图 7.2-9

(2) 单击"System Config"按钮，在弹出的设置界面中，按图 7.2-10 所示界面设置。其中单击"MD1_DB"按钮，在弹出的文件选择框中，需要选择 catcher 中加载的那个数据库文件：BPLGUInfoCustomAppSrcP_MT2502_S00_FK001_V0_01。设置完成后，单击"Save"按钮。

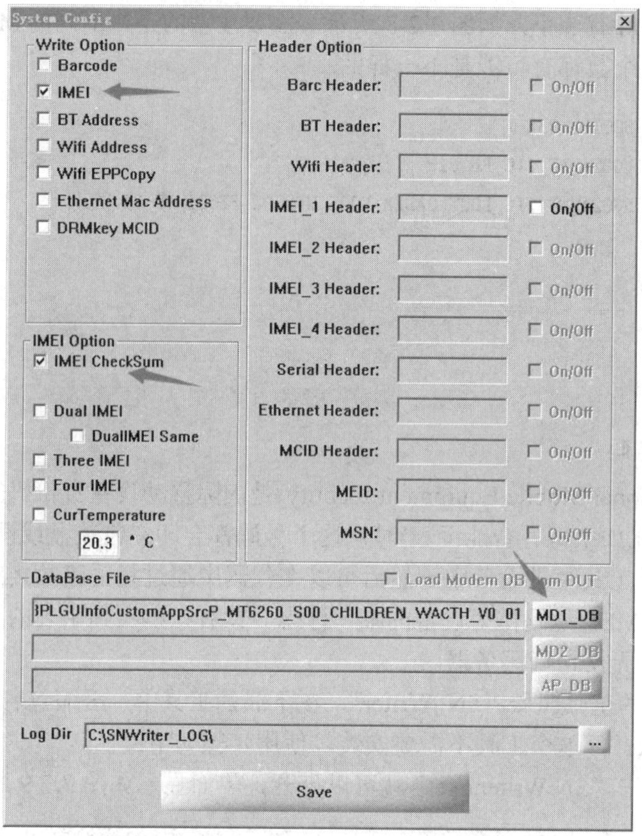

图 7.2-10

(3) 单击 "Target Type" 下拉选择框，选择 "Feature Phone"，再单击 "Start" 按钮，在弹出的对话框中输入要写入的 IMEI 号码，这个 IMEI 号必须是自己购买手表时印在包装盒上的 IMEI 号码，不能随便输入。如图 7.2-11 所示。

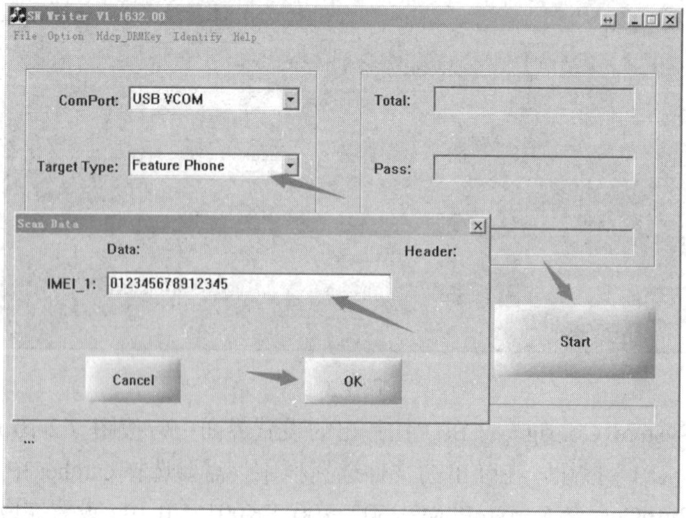

图 7.2-11

(4) 单击"OK"按钮，会出现如图 7.2-12 所示界面，此时先把设备关机，然后用 USB 线连接电脑，与烧录版本时的步骤是一样的。

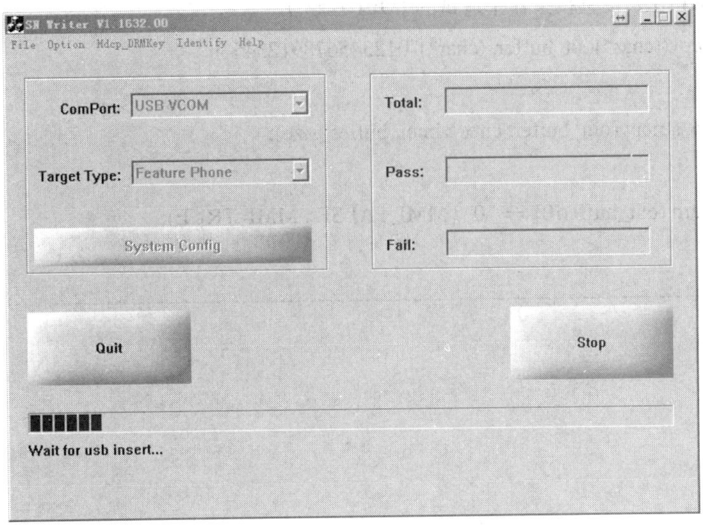

图 7.2-12

写成功之后，就会出现如图 7.2-13 所示的界面。

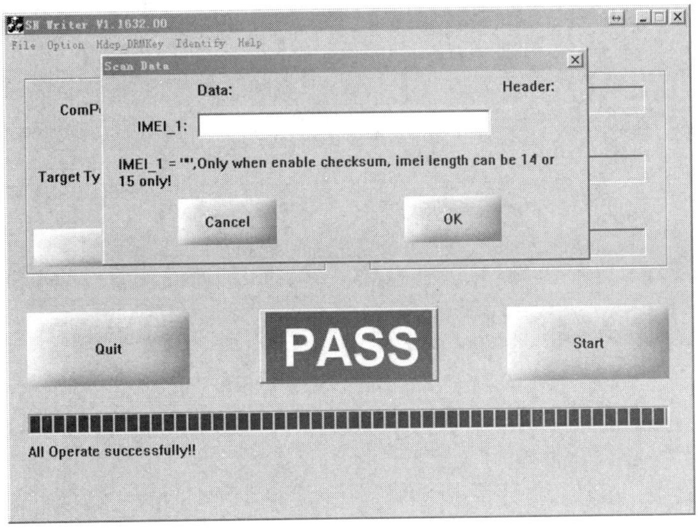

图 7.2-13

IMEI 号通常只要写入一次就可以了，但是在下载版本的时候，如果选择了格式化 (Format FAT) 下载，则需要重新写入。如果在模拟器上调试网络部分的代码，则可以在获取 IMEI 的函数接口 srv_imei_get_imei 中(plutommi\Service\ImeiSrv\ImeiSrv.c)，手动输入 IMEI 号码，代码如下(代码清单 7.2-5)：

---代码清单 7.2-5---

/*省略部分默认代码*/
MMI_BOOL srv_imei_get_imei(mmi_sim_enum sim, CHAR *out_buffer, U32 buffer_size)

```c
{
    /*省略部分默认代码*/
#ifdef WIN32/*设置模拟器中调试的 IMEI 号码*/
    strcpy((char*)out_buffer, (char*)"012345678912345");
#else
    app_strlcpy(out_buffer, cntx->imei, buffer_size);
#endif
    return (out_buffer[0] == '\0' ? MMI_FALSE : MMI_TRUE);
}
```

附录 A Source Insight 工具介绍

Source Insight 是一个面向项目开发的程序编辑器和代码浏览器,它拥有内置的对 C/C++、C# 和 Java 等程序的分析,能分析源代码并在工作的同时动态维护它自己的符号数据库,并自动显示有用的上下文信息。

1. 创建项目

用 Source Insight 创建项目步骤如下:

(1) 打开 Source Insight,选择"Project"→"New Project",在弹出的对话框中输入项目名称(比如 MT6260M)和项目文件保存的路径(笔者存放在 E:\MT6260M 中),单击"OK"按钮,如图 A-1 所示。

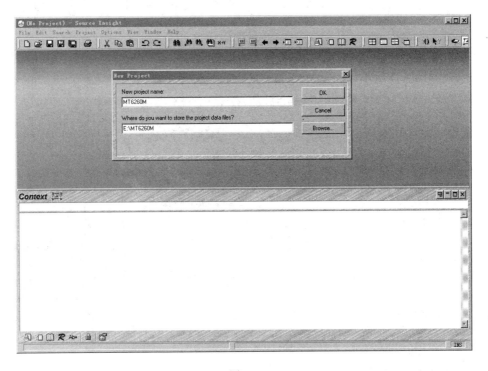

图 A-1

(2) 在弹出的对话框中再单击"OK"按钮,在"Add and Remove Project Files"对话框中,加载项目文件,选择 MT6260m,再选择右侧的"Main",然后单击"Add Tree"按钮,如图 A-2 所示。

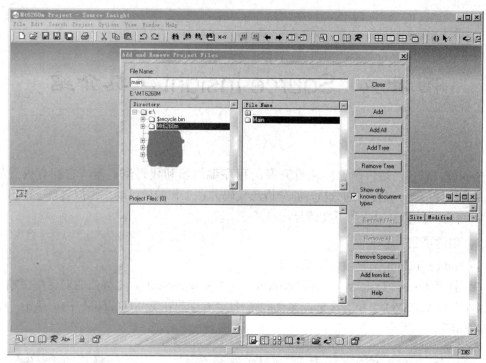

图 A-2

此时会出现一个进度条,进度条加载完成后,会弹出一个对话框,提示总共加载了多少个文件,如图 A-3 所示,提示为 18233 个文件。

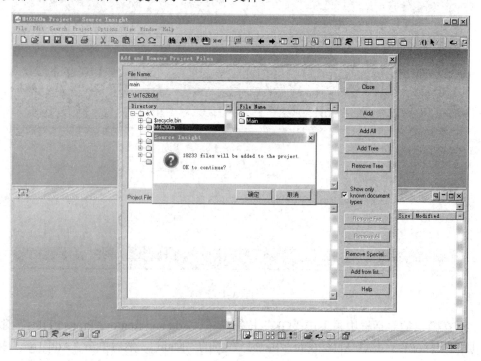

图 A-3

(3) 点击"确定"按钮，然后在"Add and Remove Project Files"窗口中单击"Close"按钮。到这里 Source Insight 工程就建好了。但是在新创建的工程中，代码文件中的很多变量和函数显示都是灰色的，那是因为代码符号数据库没有建立。在"Project"→"Rebuild Porject"对话框中选择"Re-Parse all source files"选项，单击"OK"按钮，如图 A-4 所示。进度条加载完成后，代码文件中的变量和函数就会显示为彩色了。

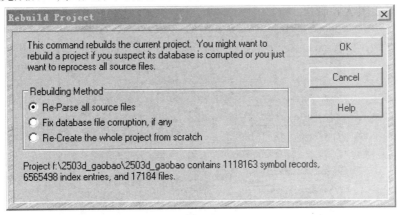

图 A-4

2．常用设置

（1）选择背景色。选择"Options"→"Preferences"→"Colors"，在弹出的对话框中选择"Window Background"选项。白色的背景看久了眼睛容易疲劳，对视力保护比较在意的人，可以调成护眼色，即色调：85；饱和度：123；亮度：205；红：199；绿：237；蓝：204。如图 A-5 所示。

图 A-5

(2) 设置字体。

字体在"Options"→"Document Options"对话框中可以修改。Source Insight 默认字体是 Verdana，这个字体下的字符是不等宽的。比如"llllllllll"和"MMMMMMMMMM"，同样 10 个字符，但显示的长度却相差很多。用 Verdana 来看程序，使得本应对齐的代码就歪了。解决的方法是使用等宽的字体，比较推荐的是用 Courier New，字体大小可根据自己的习惯设置。如图 A-6 所示。

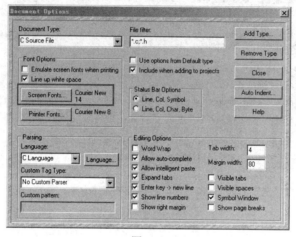

图 A-6

(3) SourceInsight 标题栏显示完整路径。

选择"Options"→"Preferences"→"Display"，在弹出的对话框中去掉"Trim long path names with ellipses"选项前面的勾，如图 A-7 所示。

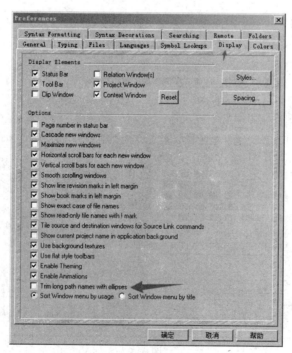

图 A-7

(4) 解决 Tab 键缩进问题。

在"Options"→"Document Options"对话框的右下角"Editing Options"栏里，勾选"Expand tabs"项，"Tab width"设置为 4(通常是默认的)。设置完成后直接单击"Close"按钮关闭，会自动保存，现在 Tab 键的缩进和四个空格的缩进在 Source Insight 里面看起来就对齐了，如图 A-8 所示。

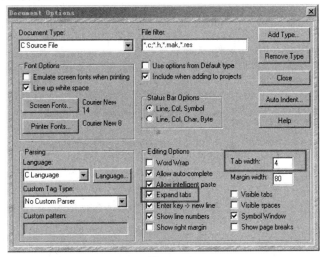

图 A-8

(5) Source Insight 中代码自动缩进对齐。

在 C 程序里，如果遇到行末没有分号的语句，如 if、while、witch 等，写到该行末按回车键，则新行自动相对上一行缩进一个 Tab 键，但如果输入的中括号是"{}"则会与上一行对齐。

在"Option"→"Document option"对话框中的"Auto Indenting"下，"Auto Indent Type"有三种类型 None、Simple、Smart，选择"Simple"类型，并把"Smart Indent Options"下的两个复选框全去掉，如图 A-9 所示。

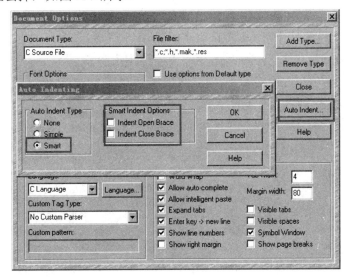

图 A-9

(6) 在项目中添加特定类型的文件(文件类型过滤器)。

C 语言的代码文件只有"*.c"和"*.h"文件,但我们在使用 Source Insigh 管理 MTK 代码时,还需要编辑"*.res"文件和"*.mak"文件,所以要把这两类文件也添加到 C 语言文件过滤器中。在"Options"→"Document Options"对话框中,点击左上角的"Document Type"下拉菜单,选择"C Source File",然后在右边的 File Filter 里"*.c, *.h"的后面加上", *.mak, *.res"(文件类型要用英文逗号隔开),接着单击"Close"按钮就可以了,如图 A-10 所示。

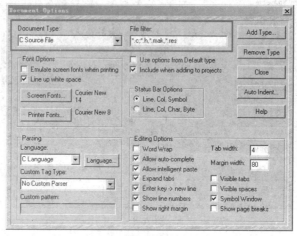

图 A-10

需要注意的是,加入新的文件类型后,需要再次执行 Project→Add and Remove Project Files 中的 Add Tree 操作,才能把这些新类型的文件添加到项目中。

(7) 解决中文显示乱码问题。

Source Insight 中默认的字体是不支持中文的,我们在代码中写的中文注释往往不能正常显示,或者显示得很难看。

在"Option"→"Style Properties"对话框的左侧 Style Name 中选择"Comment",然后在 Font Name 中选择"Pick...",在弹出的字体选择对话框中,分别选择"宋体""常规""小四",然后单击"确定""Done"按钮,如图 A-11 所示。

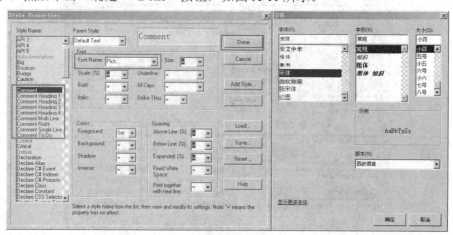

图 A-11

字体设置好后，Font Name 处就会显示"宋体"，此时代码注释中的中文字符应该显示正常了，如果还有问题，就把 Style Name 中以为"Comment"开头的几个字体全部设置成宋体。

(8) 设置宏的开关条件。

选中某个宏，鼠标右键单击，在右键菜单中选择"Edit Condition..."，弹出如图 A-12 所示对话框。

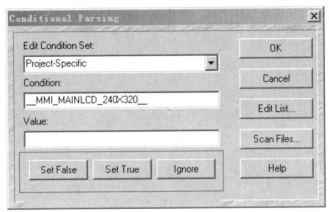

图 A-12

图中有三个按钮，如果已知这个宏没有定义，则单击"Set False"按钮，把宏的 Value 置为 0，再单击"OK"按钮，则被这个宏所包含的代码就会显示为灰色；同样方法，如果已知宏已被定义，则单击"Set True"，把宏的 Value 置为 1；如果不知道当前宏的值，则选择"Ignore"，把宏的 Value 清空，这也是宏的默认状态。另外如果宏的值不为 0 也不为 1，而是其他的数值，比如定义屏幕宽度的宏，值为 240，则可以直接在 Value 输入框中输入 240，再单击"OK"按钮。

3．常用快捷键

熟练使用快捷键，能够提高 Source Insight 的操作速度，从而提高工作效率。以下列出几个常用的快捷键。

(1) Ctrl+0：鼠标光标定位到文件检索输入框中，常用于打开文件。

(2) ALT+L：鼠标光标定位到右侧文件结构检索输入框，常用于快速查找文件中的全局字符，比如函数或全局变量、宏定义。

(3) Ctrl+A：如果编辑了多个文件未保存，则会保存所有文件。

(4) Ctrl+S：保存当前文件。

(5) Ctrl+/：在整个工程中的所有文件里，全局搜索符号。

(6) Ctrl+F：查找，可以设置大小写匹配和全字符匹配。

(7) shift+F8：高亮或取消高亮光标所在的字符内容。

(8) F12：增量查找(Incremental Search)。按下 F12 键后，直接在代码文件中输入想要查找的内容(仅限英文字符)，Source Insight 就可以自动匹配，有点类似 Windows 系统中通过按键快速定位文件或文件夹的功能。按下 F3 或者 F4 或者 Esc 键退出增量查找。这个功能比 Ctrl+F 的查找功能好用很多。

(9) F3：查找上一个，即往上查找。
(10) F4：查找下一个，即往下查找。
(11) F5/Ctrl+G：打开行数输入框，把光标快速定位到指定行数。
(12) F7：打开全局项目符号框，查找项目中的全局符号。
(13) F8：打开局部项目符号框，查询当前文件中的符号。
(14) F9：光标所在代码行往左移一个 Tab。
(15) F10：光标所在代码行往右移一个 Tab。通常与 F9 配合，用于代码的缩进调整。

附录 B　Beyond Compare 工具介绍

Beyond Compare 是一个文件比较工具，主要用途是对比两个文件夹或者文件，并将差异以颜色标示。比较范围包括目录、文档内容等。

1．文件和文件夹比较

以 MTK 源码为例，我们比较 Main 目录与章节源码中的"6、新增功能模块"文件夹中代码的差异。首先鼠标右键单击 Main 目录，选择"选择左侧文件夹来比较(L)"选项。如图 B-1 所示。

图 B-1

如果在右键菜单中没有发现该选项，则先通过 Beyond Compare 快捷方式打开，然后选择"工具"→"选项"，在启动项里面，勾选"在资源管理器上下文菜单中包含 Beyond Compare"，如图 B-2 所示。这个功能在安装的时候，一般会默认选上。

图 B-2

选择完左侧比较内容后，进入章节源码目录中，右键单击"6、新增功能模块"，然后选择"和"Main"比较"菜单，如图 B-3 所示。

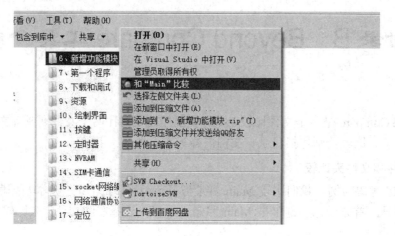

图 B-3

选择完成后，就会自动打开 Beyond Compare 工具，并且已经加载好了需要比较的内容，如图 B-4 所示。

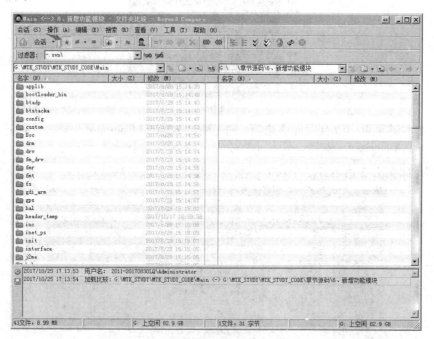

图 B-4

在图 B-4 所示界面中，我们需要了解一下顶部工具栏中一些按钮的使用，例如，当点击"显示所有"按钮时(图 B-4 中箭头所示)，会列出文件或文件夹中的所有内容，效果如图 B-4 所示。在"显示所有"按钮右边，是显示差异项的按钮，单击右侧的小箭头，里面有各种差异显示选项，如图 B-5 所示。读者可依次点击不同的按钮看看效果。下面介绍几个常用按钮。

图 B-5

(1) 显示差异项：列出所有有差异的内容，其中包括独立的内容，有差异的内容用红色标记。

(2) 不显示孤立项：过滤掉所有孤立的文件夹或文件，只显示左右两侧都有的文件或文件夹。

(3) 显示差异项但不包括孤立项：显示左右两侧都有的并且有差异的文件或文件夹。

(4) 显示孤立项：只显示左右两侧孤立的文件或文件夹。左右两侧都有的文件或文件夹全部过滤掉，不管是否有差异。

后面还有 6 个按钮，就不再一一解释了。在比较上述两个文件夹时，我们先选择"显示差异项但不包括孤立项"可以列出我们修改的系统源文件。如图 B-6 所示。

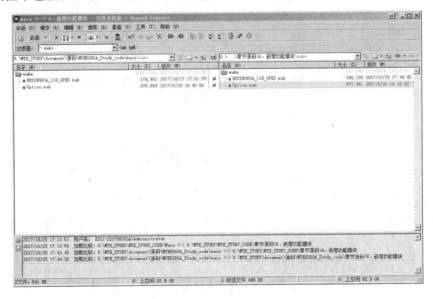

图 B-6

然后再选择"显示右侧孤立项",可以列出我们新增的文件,如图 B-7 所示。

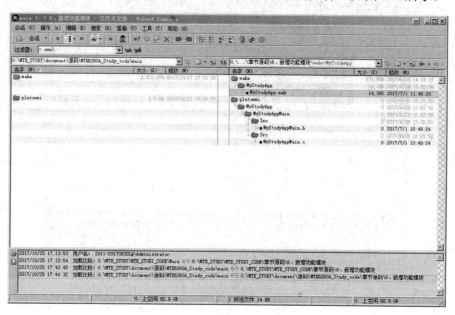

图 B-7

对于一些不需要比较的文件或文件夹,我们可以在过滤器中把它们过滤掉,点击"文件过滤器"按钮,如图 B-8 箭头所指,在弹出的对话框中,我们可以设置包含文件、排除文件、包含文件夹、排除文件夹,如图 B-8 所示。

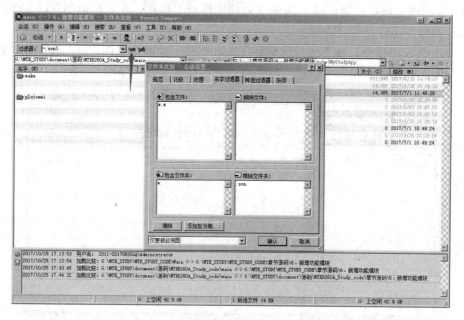

图 B-8

比如我们只比较 MTK 的代码源文件(*.c、*.h)、资源文件(*.res)、makefile 文件(*.mak),排除 svn 版本控制文件夹和 build 文件夹,则设置内容如图 B-9 所示。

图 B-9

2. 文件内容比较

选择比较内容，同文件夹比较一样，先找到要比较的第一个文件，通过右键单击，选择"选择左侧文件夹来比较(L)"选项，再找到第二个要比较的文件，右键单击，选择"和 XXX 比较"，就可以打开 Beyond Compare 文件内容比较窗口了。也可以在比较文件夹的时候，通过双击文件名打开文件内容比较窗口。在上述两个文件夹比较的内容中，我们双击打开 Make 目录下的 Option.mak 文件，显示界面如图 B-10 所示。注意在箭头指向的地方选择正确的文件编码格式，否则可能显示乱码。

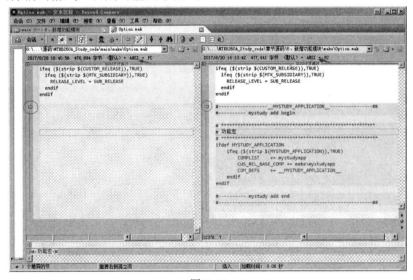

图 B-10

单击左侧圆圈标记的箭头，可以把内容同步到右侧，其实就是把右侧文件中红色标记的内容删除了。单击右侧圆圈标记的箭头，可以把内容同步到左侧，就是把右侧红色标记的内容新增到左侧文件中。也可以把光标定位到差异处，通过快捷键 **Ctrl+R** 或 **Ctrl+L** 把内容同步到右边或左边。

工具栏上的其他标签，与文件夹比较类似，鼠标停留在上面都会有字符提示。

参 考 文 献

[1] [美]凯尼格，等. C 陷阱与缺陷. 高巍，等，译. 北京：人民邮电出版社，2008.
[2] [美]Peter van der Linden. C 专家编程(英文版). 北京：人民邮电出版社，2013.
[3] [美]克尼汉，等. C 程序设计语言. 徐宝文，等，译. 北京：机械工业出版社，2004.
[4] [美] Weiss. M. A，等. 数据结构与算法分析. 冯舜玺，等，译. 北京：机械工业出版社，2004.
[5] Donahoo J, Calvert L. TCP/IP Sockets in C. 2nd ed. Practical Guide for Programmers. 2nd ed. San Francisco: Morgan Kaufmann, 2009.
[6] 陈智鹏. 走出山寨：MTK 芯片开发指南. 北京：人民邮电出版社，2010.